T0238989

TECHNOLOGY, INNOVATION and POLICY

Series of the Fraunhofer Institute
for Systems and Innovation Research (ISI)

Volume 1:
Kerstin Cuhls, Terutaka Kuwahara
**Outlook for Japanese and German
Future Technology**
1994. ISBN 3-7908-0800-8

Volume 2:
Guido Reger, Stefan Kuhlmann
**European Technology Policy
in Germany**
1995. ISBN 3-7908-0826-1

Volume 3:
Guido Reger, Ulrich Schmoch (Eds.)
**Organisation of Science and Technology
at the Watershed**
1996. ISBN 3-7908-0910-1

Volume 4:
Oliver Pfirrmann, Udo Wupperfeld and
Joshua Lerner
**Venture Capital and New Technology
Based Firms**
1997. ISBN 3-7908-0968-3

Volume 5:
Knut Koschatzky (Ed.)
**Technology-Based Firms
in the Innovation Process**
1997. ISBN 3-7908-1021-5

Volume 6:
Frieder Meyer-Krahmer (Ed.)
**Innovation and Sustainable
Development**
1998. ISBN 3-7908-1038-X

Volume 7:
Ulrike Bross, Annamária Inzelt
and Thomas Reiß
Bio-Technology Audit in Hungary
1998. ISBN 3-7908-1092-4

Volume 8:
Gunter Lay, Philip Shapira
and Jürgen Wengel (Eds.)
Innovation in Production
1999. ISBN 3-7908-1140-8

Volume 9:
Frieder Meyer-Krahmer (Ed.)
**Globalisation of R&D
and Technology Markets**
1999. ISBN 3-7908-1175-0

Volume 10:
Klaus Menrad et al.
**Future Impacts of Biotechnology
on Agriculture, Food Production
and Food Processing**
1999. ISBN 3-7908-1215-3

TECHNOLOGY, INNOVATION and POLICY 11

Series of the Fraunhofer Institute
for Systems and Innovation Research (ISI)

Emmanuel Muller

Innovation Interactions between Knowledge-Intensive Business Services and Small and Medium-Sized Enterprises

An Analysis in Terms
of Evolution, Knowledge and Territories

With 42 Figures and 19 Tables

Physica-Verlag

A Springer-Verlag Company

Dr. Emmanuel Muller
Fraunhofer Institute for Systems and Innovation
Research (ISI)
Breslauer Straße 48
76139 Karlsruhe
Germany
E-mail: em@isi.fhg.de
Internet: www.isi.fhg.de

ISSN 1431-9667
ISBN 3-7908-1362-1 Physica-Verlag Heidelberg New York

Die Deutsche Bibliothek – CIP-Einheitsaufnahme
Muller; Emmanuel: Innovation interactions between knowledge intensive business services and
small and medium sized enterprises: an analysis in terms of evolution, knowledge and territories;
with 19 tables / Emmanuel Muller. – Heidelberg; New York: Physica-Verl., 2001
    (Technology, innovation and policy; 11)
    ISBN 3-7908-1362-1

Physica-Verlag Heidelberg New York
a member of BertelsmannSpringer Science+Business Media GmbH

© Physica-Verlag Heidelberg 2001
Printed in Germany

Softcover design: Erich Kirchner, Heidelberg
SPIN 10790089   88/2202-5 4 3 2 1 0 – Printed on acid-free paper

Acknowledgements

The present work, originally constituted as a Ph.D. dissertation, would not have been possible without the constant intellectual and moral support provided over the years by Jean-Alain Héraud and the co-direction by Knut Koschatzky. I am deeply grateful for their trust and for having profoundly affected my vision of science, Europe and the future. I think that the German term of *Doktorvater* depicts perfectly how much I am indebted to them both.

The final stage of this analysis profited greatly from the attention, suggestions and kindness of Peter Wood during a sabbatical stay at the University College London in 1998. I am particularly grateful to Benoît Godin, Yvon Martineau and the INRS-Urbanisation for the generous invitation to spend several months in Montréal in 1996. It allowed me to develop an important part of the theoretical reflections underlying the investigation. I am thankful also for the opportunity offered, before starting the Ph.D., by Imre Hronsky and supported by Jean Lachmann and the Regional Council of Alsace, to carry out analyses about innovation systems in 1993 at the Technical University of Budapest. It gave me the motivation for future research.

I gratefully acknowledge my colleagues at Fraunhofer Institute for Systems and Innovation Research (ISI), Karlsruhe, particularly Christine Schädel (for friendly support and layout), Helga Traxel (for motherly attention and data treatments), and Ralf Schneider (for constant good mood and graphics).

The Ph.D. benefited from the opportunity to participate in the programme "Technological Change and Regional Development in Europe" granted by the German Research Association and in the TSER project TIPIK ("Technology and Infrastructures Policy in the Knowledge-based Economy") financed by the DG XII. To proceed forward despite the differences existing between the French and German contexts has been made possible for me thanks particularly to Lionel Pilorget, Frieder Meyer-Krahmer and Patrick Cohendet. The work also benefited from discussions with Christiane Hipp and Günter Walter.

It has been my good fortune to get so much help, encouragement and friendship at decisive times. In this respect, I wish especially to thank Alain, Angela, Béatrice, Cynthia, Frank, Herbert, Isabel, Jean-Luc, Letizia, Maria, Niels, Noemí, Oliver, Olivier, Roland, Uli, Ute, Valérie, Viola and Volker.

Through her presence, Natascha illuminated the final stage of the writing, brushing away doubts and sorrows.

Even more than for her many helpful suggestions, her infallible support and the pleasure I have to work with her, I am profoundly grateful to my colleague and friend Andrea Zenker for showing so much understanding, humour and patience.

Karlsruhe, July 2000

Contents   Page

General introduction 1

Chapter 1: Innovation as the expression of firms' evolution capacity 5

Introduction...5

1.1   Innovation as an evolutionary process 5

1.1.1   Schumpeter and the "process of creative destruction".....................5

1.1.2   Linear vs. interactive model of innovation 8

1.1.3   Innovation, evolution and firms' performance 11

1.2   Innovation as a knowledge-based process 15

1.2.1   Taking the firm as an information system..16

1.2.2   Information and the knowledge base of the firm 18

1.2.3   Knowledge-base of the firm and innovation....................................19

1.3   Nature and forms of innovation: towards a continuum
              approach...21

1.3.1   A good/service continuum?...21

1.3.2   Innovations of product/services *vs.* process innovations 23

1.3.3   A broader conception of innovation?...24

Conclusion 25

**Chapter 2: Interactions between KIBS and SMEs and impact on
innovation capacities** 27

Introduction...27

2.1   Interactions: between market and hierarchy?...................................27

2.1.1   The "make or buy" approach...27

2.1.2   The transaction costs theory 28

2.1.3   The network analysis 30

2.2   The impact of interactions with KIBS on SMEs..............................35

2.2.1   A characterisation of SMEs knowledge-base and
            innovation activities..35

2.2.2   KIBS as complementary innovation assets for SMEs 37

2.2.3   KIBS as co-innovators...39

2.3   KIBS innovation capacity and the influence of interactions
           with SMEs 42

2.3.1   A context of socio-economic and technological changes 42

2.3.2   Exploring the knowledge-base of KIBS..44

2.3.3   The impacts of interactions with SMEs on KIBS 45

Conclusion 47

**Chapter 3: Territorial determinants and evolution capacities of SMEs
and KIBS...49**

Introduction...49

3.1   Proximity and innovation 49

3.1.1   Knowledge spillovers and proximity 49

3.1.2   Proximity, accessibility of information and learning.....................50

3.1.3   Conceptualisation of proximity...51

3.2   Territory and innovation...53

3.2.1   The innovation environment and the debated relevance of
           the territory 53

3.2.2   From industrial districts to learning regions 57

3.2.3   The systemic approach 58

3.3   The impact of territorial determinants on SMEs and KIBS
           interactions and evolution 60

3.3.1   The specific impact of proximity-based interactions
           between SMEs and KIBS 61

3.3.2   The influence of the type of regional environment on the
           evolution capacities of SMEs and KIBS 63

3.3.3   The national innovation system as a determinant of SMEs
           and KIBS behaviour...65

Conclusion 66

Contents

Chapter 4: Operationalisation of the analysis **67**

Introduction...67

4.1   From the hypotheses to the key variables 67

4.1.1   The hypotheses to be tested..67

4.1.2   The conceptual model..68

4.1.3   The variables of the analysis 69

4.2   Structure of the data..72

4.2.1   The surveyed regions...73

4.2.2   The SME sample 74

4.2.3   The KIBS sample..78

4.3   The statistical exploitation procedure 82

4.3.1   The exploratory stage: the segmentation analysis (CHAID) 83

4.3.2   The middle stage: multiple correspondence analysis.....................84

4.3.3   The final stage: path-modelling..86

Conclusion 88

Chapter 5: Statistical exploitation of the SME sample...**89**

Introduction...89

5.1   Segmentation procedures..89

5.1.1   First segmentation...90

5.1.2   Second segmentation 91

5.1.3   Third segmentation 94

5.2   Multiple correspondence analysis 96

5.2.1   Contribution to variance explanation 96

5.2.2   The main dimensions of the multiple correspondence
         analysis 96

5.2.3   Interpretation of the correspondence analysis...............99

5.3   Path modelling...100

5.3.1   Variables reduction...100

5.3.2   Results...101

5.3.3   Interpretation...104

Conclusion 105

Chapter 6: Statistical exploitation of the KIBS sample 107

Introduction...107

6.1   Segmentation procedures...107

6.1.1   First segmentation..107

6.1.2   Second segmentation 109

6.1.3   Third segmentation 112

6.2   Multiple correspondence analysis 114

6.2.1   Contribution to variance explanation 114

6.2.2   The main dimensions of the multiple correspondence
           analysis 114

6.2.3   Interpretation of the correspondence analysis...............................117

6.3   Path modelling...118

6.3.1   Variable reduction 118

6.3.2   Results..119

6.3.3   Interpretation...121

Conclusion 122

Chapter 7: Main findings and policy implications 125

Introduction...125

7.1   Interpretation of the key findings 125

7.1.1   Innovation and evolution: what can be learnt from SMEs
           and KIBS?...126

7.1.2   The virtuous circle: between contamination and symbioses.........127

7.1.3   The territorial component: does space really matter? 129

7.2   Towards an integrated typology of innovation interactions 131

7.2.1   KIBS demand and supply response: Wood's model 131

7.2.2   An integrated typology of knowledge exchanges 134

7.2.3   Examples...137

7.3   Implications for policies 142

7.3.1   Innovation: rather a matter of knowledge than of
           technique?...142

7.3.2   The concept of induced support 143

7.3.3   Some elements contributing to a renewed regional policy
           agenda..145

Conclusion 148

General conclusion...149

References..153

Appendix 169

    Appendix A: Basic frequencies of the selected variables...............................171

    Appendix B: Selected bivariate tests...183

    Appendix C: PROBIT analysis 191

List of figures

Figure 1.1:   The *Schumpeter mark I* model..7

Figure 1.2:   The *Schumpeter mark II* model 8

Figure 1.3:   The conventional linear model 9

Figure 1.4:   The chain-linked model 10

Figure 1.5:   The virtuous circle and the vicious circle..15

Figure 1.6:   The firm as an information system according to Davis
               (1974)...16

Figure 1.7:   The firm as an information system according to Le Moigne
               (1986)...17

Figure 1.8:   A context of expanding knowledge 19

Figure 1.9:   The knowledge pyramid 20

Figure 1.10:   The tangible-intangible dominant continuum...23

Figure 2.1:   Possible activities of Institutions of Technological
               Infrastructure (ITI)...34

Figure 2.2:   The scientific and technical information system of a SME.............37

Figure 2.3:   *Schumpeter mark III* derived from *Schumpeter mark I*....................40

Figure 2.4:   *Schumpeter mark III* derived from *Schumpeter mark II* 40

Figure 2.5:   The contribution of KIBS to the innovation capacity of
               SMEs..41

Figure 2.6:   The place of KIBS in a perspective of socio-economic and
               technological changes..43

Figure 2.7:   The production process leading to the service output......................45

Figure 2.8:   The contribution of SMEs to the innovation capacity of
               KIBS...47

Figure 3.1:   The distance decay function 52

Figure 3.2:   Networking and the external environment of the firm.....................54

Figure 3.3:   The model of immaterial infrastructure of a territory 56

Figure 3.4:   The mutual contribution of KIBS and SMEs to their
               innovation capacity: what are the territorial determinants? 61

Figure 4.1:   The conceptual model 69

Figure 4.2:   The surveyed regions 74

Figure 4.3:   The "stat-mix" procedure...82

Figure 4.4:   Example of CHAID tree-diagram...84

Figure 4.5:   Canonical analysis, $\xi \in W_1$ and $\eta \in W_2$ presenting a
                minimal angle..86

Figure 4.6:   Path modelling – dependent and explanatory variables 87

Figure 4.7:   Path modelling – direct and indirect dependencies 88

Figure 5.1:   "Growth" as a dependent variable 91

Figure 5.2:   "Introduction of innovation" as a dependent variable 93

Figure 5.3:   "Interaction with KIBS" as a dependent variable...............................95

Figure 5.4:   Multiple correspondence analysis of the SMEs sample....................98

Figure 5.5:   Path modelling: the SMEs sample..103

Figure 6.1:   "Growth" as a dependent variable 109

Figure 6.2:   "Performance of innovation" as a dependent variable 111

Figure 6.3:   "Interaction with SMEs" as a dependent variable 113

Figure 6.4:   Multiple correspondence analysis of the KIBS sample...................116

Figure 6.5:   Path modelling: the KIBS sample..120

Figure 7.1:   KIBS demand and supply response from a spatial
                perspective 133

Figure 7.2:   The wheel of knowledge interactions implying KIBS and
                SMEs...135

Figure 7.3:   The "holy trinity" of regional economics according to
                Storper..146

List of tables

Table 4.1:   Overall structure of the variables...70

Table 4.2   The activities covered by the SME sample 75

Table 4.3:   The seven aggregated SME sectors 76

Table 4.4.:   The sector distribution of each regional sample.............................77

Table 4.5:   Variables extracted from the SMEs survey......................................78

Table 4.6:   The activities covered by the KIBS sample.......................................79

Table 4.7:   The four aggregated KIBS sectors..80

Table 4.8:   The sector distribution of each regional sample.............................80

Table 4.9:   Variables extracted from the KIBS survey..81

Table 5.1:   Dependent and explicative variables of the CHAID
            procedures 89

Table 5.2:   Eigenvalues of the correspondence analysis 96

Table 5.3:   Discrimination measures of the correspondence analysis.................99

Table 5.4:   Set of dichotomic variables used for the path-modelling................101

Table 5.5:   Results of the PROBIT analysis 102

Table 6.1:   Dependent and explicative variables of the CHAID
            procedures 107

Table 6.2:   Eigenvalues of the correspondence analysis 114

Table 6.3:   Discrimination measures of the correspondence analysis...............115

Table 6.4:   Set of dichotomic variables used for the path-modelling................118

Table 6.5:   Results of the PROBIT analysis 119

List of acronyms and abbreviations

CHAID:	Chi-squared Automatic Interaction Detector
ECU:	European Currency Unit
GREMI:	*Groupe de Recherche Européen sur les Milieux Innovateurs*
HTSF:	High Technology Small Firms
IT:	Information Technologies
ITI:	Institutions of Technological Infrastructure
KECU:	Kilo ECU
KIBS:	Knowledge-Intensive Business Services
MSA:	Metropolitan Statistical Area
NIS:	National Innovation System
OCDE:	*Organisation de Coopération et de Développement Économique*
OECD:	Organisation for Economic Co-operation and Development
PROBIT:	Maximum-likelihood probit estimation
RIS:	Regional innovation system
RTE:	*Réseau technico-économique*
SME:	Small and medium-sized enterprise

General introduction

The current decade has stressed the reinforcement of three major trends affecting developed economies since at least half a century. The first trend concerns the importance of innovations for firms, regions and nations. Even if innovation, however defined, is not a recent phenomenon, the crucial role innovativeness plays in a world of increased competition reveals its importance as a major economic phenomenon. A second prominent trend corresponds to the "tertiarisation of the economy", *i.e.* the considerable and continuous expansion of the share of services in contemporary economies.[1] The third trend consists in the substantial changes affecting the economical meaning of "space". Signs of these substantial changes can be found in the increased unity of "space" (the globalisation phenomenon), in the appearance of new territories (as suggested by the emergence of a "virtual continent"[2]) as well as in the reconsideration of existing ones (for instance, induced by new flexible production systems). As such, these three trends express a triple breakdown: a breakdown in the *time scale* of the economy (due to the increasing velocity of economic activities), a breakdown in the *nature* of the economy (since "intangible production" is expanding) and a breakdown in the *space(s)* of the economy (because of shifting dimensions). Certain categories of activities or actors may highlight and even symbolise this triple rupture in diverse ways. An illustration can be found in the dominant market position of multinational corporations or "global enterprises" that has never been reached before. In parallel, these ruptures particularly highlight the development of what has been pointed out as the "symbolic workers' high-tech nomadism".[3] It also explains the (controversial) feeling that a quasi-instantaneous and virtual circulation of almost everything may be possible. The present work deals with activities and actors deeply affected by this triple rupture: the analysis focuses on innovation interactions taking place between small and medium-sized manufacturing enterprises (SMEs) and knowledge-intensive business services (KIBS).

The objective of the analysis is to investigate the meaning, nature and consequences of innovation interactions between SMEs and KIBS. Small and medium-sized manufacturing enterprises (SMEs) can be identified according to their size (SMEs

---

1   In this respect, *cf.* for instance Illeris (1996).

2   This metaphor, borrowed from Attali (1998), refers not only to the web as such but more generally to the consequences of the recent boom of virtual digitalised applications, and literally, to the emergence of "new economic worlds".

3   There are different possible meanings given to these terms: *cf.* for instance the concept of "symbolic analyst" as described by Reich (1991) or of "symbolic worker" as presented by Rifkin (1996). For a controversial approach to these views and a critic of the emergence of "new economic worlds", one may refer to the contribution of Châtelet (1998).

are usually classified with reference to the 500 employees limit) and thus defined in opposition to large manufacturers. In this respect, SMEs present specific characteristics in terms of innovation activities and knowledge-related capacities.[4] In a similar way, knowledge-intensive business services (KIBS) can be described as firms performing, mainly for other firms, services encompassing a high intellectual value-added. Thus, KIBS correspond broadly to "consultancy services".[5] The concept of innovation interactions relates to the provision of, or the benefit from support affecting innovation processes.[6] In addition to the actors considered and to the nature of the phenomena examined, the particular interest of this investigation focuses on the concept of a virtuous circle linking KIBS and SMEs. In this respect, a particular emphasis is given to the question of the role and nature of the knowledge base of the firm in an evolutionary perspective. The integration of the spatial dimension strengthens the examination of the consequences of interactions between the considered firms on their respective knowledge base and on their related innovation capacities.

All these elements require theoretically and empirically-based investigation. To this end, a new data analysis methodology has been specially developed for this investigation. This methodology consists of a combination of complementary statistical procedures which, as a consequence, provides a tool allowing a reliable examination of the data collected. In fact, the samples considered in the analysis (SMEs and KIBS located in France and in Germany) constitute a source of rich, complex, and diversified information. Consequently, the investigation is at the crossroads of three fields of economics: (i) it corresponds to the tradition of economics of innovation (merging industrial economics and innovation sociology); (ii) it belongs *de facto* to the field of economics of services; and (iii) it consists of an analysis in terms of territories, encompassing as such the mutual contributions of regional economics and of economic geography.

The whole analysis concentrates on the principle of a virtuous innovation circle associating SMEs and KIBS. In this respect, the key results related to SMEs and KIBS can be synthesised according to five main points. At first, it appears in the light of the investigation, that innovation phenomena consist of complex learning processes, based on knowledge interactions and which express firms' evolutionary capacities. Secondly, it can be asserted that the interactions examined support a reinforced integration of SMEs and KIBS in their respective innovation environment

---

4   More generally, considering SMEs and innovation, see Acs and Audretsch (1990). The identification modalities related to SMEs for the tasks of the investigation are outlined in section 4.2.2.

5   A detailed description of KIBS can be found in Miles *et al.* (1994). The identification modalities related to KIBS for the tasks of the investigation are outlined in section 4.2.3.

6   Innovation interactions taking place between SMEs and KIBS will be examined in detail in sections 2.2 and 2.3.

as well as an improved activation of their internal and external innovation resources. Referring more precisely to the evolution patterns of the observed SMEs, statistical evidence showing the impact of interactions with KIBS on SMEs' innovation capacities could be established. Additionally, it can be assumed that SMEs' economic performance are at least partially determined by elements constituting their knowledge base, notably by the existence of interactions with KIBS. Considering the population of KIBS examined in parallel, strong empirical evidence could be found to support the idea of a positive impact of interactions with SMEs on KIBS' innovation capacities and economic performance. Finally, regarding the spatial dimension of the analysis it could be seen that proximity matters more when information flows from SMEs and knowledge is developed by KIBS than when information flows from KIBS and knowledge is developed by SMEs. Furthermore, the investigation revealed the statistically significant influences of national innovation systems on interaction and innovation patterns of the examined SMEs and KIBS, whereas it was not possible to establish such influences for the regional environments considered.

The analysis is divided into three main steps. The first step consists of theoretical reflections which result in a set of hypotheses. In the second step, the research is operationalised and the hypotheses put forward are tested statistically. The final step concentrates on the interpretation and implications of the key findings. The work is structured in seven chapters described below. The first chapter investigates the nature of the innovation phenomenon. The views adopted to depict evolution patterns in manufacturing and service activities, presenting innovation notably as an interactive process, correspond mainly to the evolutionary approach of the firm. The second chapter aims at the conceptualisation of innovation interactions between KIBS and SMEs. Subsequently, their potential impacts on SMEs' and KIBS' innovation capacities are reviewed and lead to the belief in a virtuous circle or "co-evolution" associating the innovation activities of the studied firms. Chapter 3 introduces the spatial dimensions in the analysis. More precisely, it attempts to examine the influence of proximity, regional and national environment on SMEs' and KIBS' evolution patterns. Following these theoretical reflections, the fourth chapter provides the operationalisation of the analysis. To this end, a conceptual model synthesising the developed hypotheses is proposed, the structure of the collected data is presented and the statistical tools selected for the empirical investigation are described. Chapters 5 and 6 deal with the empirical investigation of the SME and the KIBS samples respectively. The statistical findings resulting from the specific combination of three different exploitation procedures provide empirical support to the belief in a virtuous circle linking SMEs and KIBS. The final chapter reviews the key findings of the investigation and compares them to other empirical studies. Furthermore, chapter 7 proposes a typology of innovation interactions featuring knowledge exchanges involving SMEs and KIBS and, last but not least, considers the policy implications of the analysis.

Chapter 1: Innovation as the expression of firms' evolution capacity

Introduction

This chapter is devoted to the conceptual exploration of the innovation phenomenon itself. Innovation is at first presented as a process, specific to the evolution of the firm. Then, in the next section, the informational content of innovation is featured in relation with the knowledge base of the firm. Finally, and after having discussed the nature and forms of innovation in manufacturing and service firms, a continuum approach or broader conception of innovation is proposed.

1.1 Innovation as an evolutionary process

Among the literature devoted to innovation, the evolutionary or *neo-Schumpeterian* school occupies a particular place. In fact, taken as a whole, this approach concentrates on the specific characteristics of the innovation phenomenon. In this respect, the evolutionary or *neo-Schumpeterian* vision of innovation is quite different from the neo-classical one. As a consequence, it is possible to elaborate models related to the introduction of innovation, and to suggest the existence of a relationship between the propensity of a firm to innovate and its performance. To understand the contribution of evolutionary economics it is necessary to refer, at first, to Schumpeter's "process of creative destruction". Then, it is particularly relevant to confront the two main types of models depicting innovation: the linear and the interactive models. Last, but not least, to consider the consequences of innovations on the performance of firms reinforces clearly the acceptance of innovation as an evolutionary process.

1.1.1 Schumpeter and the "process of creative destruction"

According to Schumpeter (1950), the **"process of creative destruction"** rules the historical evolution of capitalism. This process re-covers five main types of innovations, which are distinguished by Schumpeter (1935): (i) new consumption objects; (ii) new production and transport methods, (iii) new markets, (iv) new sources of production materials, (v) new market position.[7] Schumpeter's vision of innovation

---

7 *Cf.* Schumpeter (1935, p.100): *"Herstellung eines neuen, d.h. dem Konsumentenkreis noch nicht vertrauten Gutes oder einer neuen Qualität eines Gutes, (...) Einführung einer neuen, d.h. dem betreffenden Industriezweig noch nicht praktisch bekannten Produktionsmethode, (...) Erschließung eines neuen Absatzmarktes, (...) Eroberung einer neuen Bezugsquelle von Rohstoffen*

goes beyond the narrow conception adopted by the standard neo-classical approach, where innovation is reduced to a simple new combination of production factors.

In the neo-classical perspective, it is more worthwhile to speak from "technology" than from "innovations". "New" productive combinations or "new" production factors are simply assumed to appear, from time to time, resulting from economically exogenous processes. In other words, in the standard neo-classical theory, the firm corresponds to a kind of **"technological black box"**[8], which is only considered in view of its market relations. The firm is seen as "technologically efficient": it applies efficient factor combinations and establishes market-relations for the purchase of inputs (resources) and the sale of outputs (production). The firm is thus continuously confronted with a set of decisions related to inputs to obtain and outputs to produce and sell. Those choices are ruled by the principle of profit maximisation (or firm value maximisation) according to exogenous conditions (*i.e.* market conditions). Technology is thus perceived as a level of combination of inputs available on the market, and technological choices are performed in a costs minimisation (or profit maximisation) logic.

Schumpeter, contrary to the standard neo-classical theory, conceives the competitive environment of firms as one made up of struggles and motions. The firm, for Schumpeter, is involved in a process of dynamic selection and not of equilibrium. The phenomenon of **"creative destruction"** constitutes the core element of this dynamic selection. Thus, the motor of economic development is provided by the **"innovator-entrepreneur"** who, according to Schumpeter, modifies existing situations. Of course, profit constitutes one of the objectives of firms and innovation may be seen as a means for growing profits.[9] Nevertheless, contrary to the standard neo-classical approach, Schumpeter does not attribute to firms a profit detailed calculation and maximisation logic.[10]

Nevertheless, the *Schumpeterian* approach to innovation is not monolithic and has evolved with time. The evolution of Schumpeter's conception of innovation may, at least partially, be summarised by the passage of the so-called **"*Schumpeter mark***

---

oder Halbfabrikaten, (...) Durchführung einer Neuorganisation wie Schaffung einer Monopolstellung (...) oder Durchbrechen eines Monopols".

8   As a reference to the "*black box*" in which Rosenberg (1982) tries to enter.

9   Particularly due to the constitution of short-time monopolies gaining advantage from the introduction of innovations.

10   The opposition stressed between neo-classical and *Schumpeterian* approaches in terms of conceptualisation of the production process is - voluntary - reduced to its simplest expression. Since the aim of this survey is to introduce the *neo-Schumpeterian* of evolutionary analysis: (i) neo-classical concepts are presented as a coherent and homogenous set; (ii) *Marxist* contributions are ignored; and (iii) the *Schumpeterian* conception is summarised on the basis of the models provided by Freeman, Clarke and Soete (1982).

I'' model to the *"Schumpeter mark II "* model (*cf.* Phillips, 1971 and Freeman, Clarke and Soete, 1982). The *Schumpeter mark I* model (*cf.* figure 1.1) corresponds to a vision of innovation taking place in a competitive or entrepreneurial capitalism, characterised by - not economically determined - inventions and exogenous scientific discoveries. The innovative entrepreneurial activity, in this context of competitive capitalism, consists mainly: (i) in identifying within the "available knowledge of inventions and discoveries" elements which bear an economical potential; and (ii) in implementing a socialisation process which transforms those elements into innovations. These two steps (identification and implementation) allow the "innovator-entrepreneur" to start the process of "creative destruction".

Figure 1.1: The *Schumpeter mark I* model

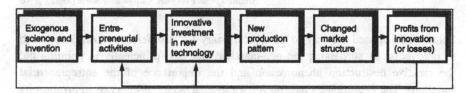

Adapted from Freeman, Clarke and Soete (1982), p. 39

Succeeding the previous one, the *Schumpeter mark II* model (*cf.* figure 1.2.) corresponds to a conceptualisation of innovation in the framework of a monopolist capitalism. This second model, as opposed to the first one, is characterised: (i) by **endogenous innovations**; and (ii) by research activities performed essentially within the **R&D departments of large firms**. This evolution corresponds to a shift from a conception focused on the individual entrepreneur as innovation source to a conception stressing the role of collective innovation functions within firms. It corresponds to a full recognition of the importance of institutionalised R&D (taking place mainly in large firms).

Figure 1.2:   The *Schumpeter mark II* model

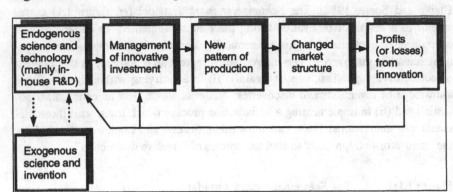

Adapted from Freeman, Clarke and Soete (1982), p. 40

Schumpeter opened the way to the evolutionary (or *neo-Schumpeterian*) approach
to innovation by surpassing the neo-classical analysis (principally in pointing out
the creative destruction phenomenon and the importance of the entrepreneurial
function). The evolutionary approach tries to analyse the process of innovation by
integrating notably interaction effects, accumulation modalities and learning routines. These aspects can be detailed in confronting linear and interactive models of
innovation.

1.1.2   Linear *vs.* interactive model of innovation

The profusion of innovation models affiliated to the evolutionary approach reveals
the complexity of innovation phenomena.[11] In the following, linear and interactive
models of innovation will only be reviewed briefly. It appears that the interactive
model (the so-called "chain-linked model") goes beyond the conventional one by
integrating feed-back loops and stressing the interactive character of innovation
processes.

---

11   For a detailed survey of innovation-related models, see for instance Schmoch *et al.*, (1993, pp.
73-100).

Figure 1.3: The conventional linear model

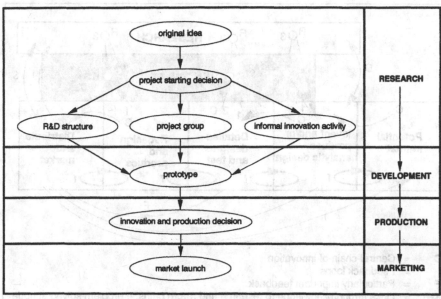

original idea

project starting decision

RESEARCH

R&D structure   project group   informal innovation activity

prototype

DEVELOPMENT

innovation and production decision

PRODUCTION

market launch

MARKETING

Source: Kline and Rosenberg (1986), p. 286, modified by Gallouj (1994), p. 194

The conventional linear model (*cf.* figure 1.3) may be seen as a prolongation of the original *Schumpeterian* modelling trial in which the only feed-back loops (if any) are the profits or losses generated by innovations (*cf.* figures 1.1 and 1.2). The linear model can also be described as *sequential* since it articulates **several distinct sequences**: (i) the original idea or the corresponding research; (ii) the development of the idea or the constitution of a prototype; (iii) the "physical translation" of the innovation in the production process; and, finally, (iv) the introduction of the innovation to the market. This quasi mechanical succession of stages, and quite strict separation between "conception "and "realisation" of innovation, are strongly rejected by Kline and Rosenberg (1986) in the frame of their "chain-linked model" (*cf.* figure 1.4).

Figure 1.4:   **The chain-linked model**

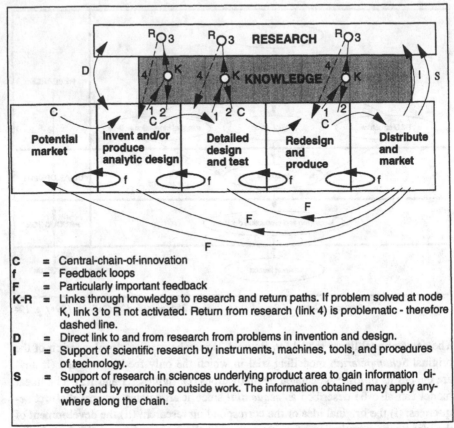

C   =   Central-chain-of-innovation
f   =   Feedback loops
F   =   Particularly important feedback
K-R   =   Links through knowledge to research and return paths. If problem solved at node
          K, link 3 to R not activated. Return from research (link 4) is problematic - therefore
          dashed line.
D   =   Direct link to and from research from problems in invention and design.
I   =   Support of scientific research by instruments, machines, tools, and procedures
          of technology.
S   =   Support of research in sciences underlying product area to gain information di-
          rectly and by monitoring outside work. The information obtained may apply any-
          where along the chain.

Source: Kline and Rosenberg (1986), p. 290

The "chain-linked model" by Kline and Rosenberg (1986) can be interpreted in terms of different levels: while the linear model features only one single level, the "chain-linked model" comprises five of them. The first level consists of the **central chain** which - extending from the "conception" to the "distribution" of innovation - can be seen as the integration of the linear model. The second level is constituted by the feed-back loops associating phases of the central chain: each stage is linked to the preceding one, and the last (marketing and distribution) to all the other phases. The third level links the central chain of innovation to research[12] which constitute a kind of "knowledge-stock" likely to stimulate each step of the innovation process (and not only its start contrary to the linear model). The last levels are less frequent and feature situations, where respectively: (i) dramatic scientific shifts generate radical innovations; and (ii) innovations support the expansion of scientific knowl-

---

12 The term *research* encompasses here internal and external activities.

edge. This approach reflects in a more realistic way the complexity of innovation
processes than the linear model. As a consequence, **the adoption of the "chain-
linked model" as reference model for the analysis of innovation-interactions
between SMEs and KIBS leads to three observations of central importance:**

(1) **the intrinsically interactive nature of innovation;**

(2) **the central role of knowledge for innovation; and**

(3) **the absence of conceptual restrictions reducing the validity of the approach to manufacturing firms only.**

These observations will influence in a critical way the next steps of the investigation since they allow a better understanding of the nature of innovation as a process.
After reviewing the impacts of innovation in terms of firms' evolution and performance (in section 1.1.3), innovation will be inspected as a knowledge-based phenomenon (section 1.2). Then, the characteristics of innovation taking place in manufacturing and in service firms respectively will be discussed (section 1.3). Finally,
the second chapter will focus on the interactive nature and on the knowledge content of innovation in exploring relationships between SMEs and KIBS.

1.1.3   Innovation, evolution and firms' performance

As it has been featured above, evolutionary economics, in presenting innovation as
a **process**, constitute a break-up with the orthodox vision. In order to better identify
the implications for the analysis of interactions between SMEs and KIBS three aspects of the evolutionary approach will be particularly detailed: (i) the type of cognitive behaviours and decision rules postulated; (ii) the essential contribution in
terms of learning effects; and (iii) the role of the selection environment and consequences in terms of firms' performance.

An important contribution of the evolutionary theory concerns firms' decision rules.
In this approach, the neo-classical hypotheses of maximisation principle and ra-
tional choices are substituted by bounded rationality in association with the notions
of routines and heuristics. Combining the behavioural approach of individual firms
proposed by Nelson and Winter (1974) with the principle of bounded rationality (*cf.*
Simon, 1982), the evolutionary vision assumes the following: (i) at each moment,
the behaviour of a firm is a function of a set of decision rules; and (ii) this set of
decision rules makes the link between a domain of environmental stimuli and a
spectrum of possible reactions. In other words, and contrary to neo-classical principles, evolutionary economics put forward "firms behaviours"[13] which are not (ex-

---

13 This can be seen as inspirited by the "behaviourist" theory (notably works like Cyert and
March, 1963) and the learning models of Arrow (1962).

clusively) ruled by profit maximisation. In the short-run, firms' decision rules are
supposed to be stable. Nevertheless, these decision-rules are not immutable and it
appears sometimes necessary for firms to modify them. In this respect, the **"search"**
and **"problem-solving"**[14] activities constitute one of the main sources of modifica-
tion of existing decision rules. Thus, according to Nelson and Winter (1974, pp.
891-894), a theoretical model of the search process should, independently of the
types of rules involved, answer the following questions:

(1) what are the goals of the search process and what is the influence of these
     goals on the process?

(2) what are the determinants of the intensity, the direction and the strategy
     of the search activity?

(3) what does the search domain consist of (in terms of decision-rules sets)?

These basic questions related to search processes provide to the present investigation relevant elements guiding the understanding of innovation from an evolutionary perspective. The following characteristics are notably attributed to innovation.
Firstly, evolutionary economics attribute a **cumulative and interactive character**
to innovation processes. The importance of accumulation and interaction appears
especially by analysing learning effects, for instance under the form of "learning by
doing" (Arrow, 1962), of "learning by using" (Rosenberg, 1976) and of "learning by
interacting" (Lundvall, 1988).[15] Secondly, in the evolutionary approach, innovation
reveals a **specific character**. The specifics of innovation abide by the principle of
historical trajectories and paradigms (Dosi, 1982) and integrate the conception of
the tacit nature of acquired knowledge. This leads to the idea of a programmed
character of innovation in the behaviour of the firm (Clark, 1986). Finally, innovation has an **institutionalised character**. The institutionalisation features mainly the
role played by the selection environment of innovation (Nelson and Winter 1974,
1975, 1977) and the importance of R&D departments (Freeman, 1982). To summarise, the evolutionary approach distinguishes itself from the standard neo-classical
theory by proposing two main conceptual elements. On the one hand, a natural
definition of the innovation phenomenon, *i.e.* changes in existing decision rules is
proposed. On the other hand, a distinct entrepreneurial function, *i.e.* the "search"
and "problem solving" activities are introduced. The following expression allows
the synthesis of the conception of innovation in the evolutionary approach: **Innova-**

---

14 The terms **"search"** and **"problem-solving"** activities may cover a broad meaning, and should
    not be reduced to formalised R&D (*cf.* Nelson and Winter, 1974, p. 982).

15 These three types of learning effects are additionally linked to firms' absorptive capacities put
    forward by Cohen and Levinthal (1989): "*Economists conventionally think of R&D as generat-
    ing one product: new information. We suggest that R&D not only generates new information,
    but also enhances the firm's ability to assimilate and exploit existing information*" (Cohen and
    Levinthal, 1989, p. 569)

tion is a non maximising, interactive, cumulative, specific and institutionalised process.[16]

The principle of the **selection environment** (as presented by Nelson and Winter,1974, pp. 891-894) is based on the hypothesis that firms applying adequate decision rules profit from their decisions (and become successful). Analogously, it is assumed that firms following inappropriate decision rules will collapse. This selection is operated by the environment.[17] Indeed, the concept of a **selection environment covers at the same time: (i) the** *medium by which* **exogenous influences are transmitted to the firm; and (ii) the** *medium through which* **firms exert a mutual influence.** Thus, the selection environment determines both influences on firms and the way those influences affect firms' decisions and behaviours. However, as underlined above, decision rules are not immutable, they evolve with time and in the course of events. The search and selection processes constitute simultaneous and interactive aspects of evolution, or in other words, it is the combination of search and selection which makes firms evolve. This allows the linking of the concepts of "selection environment" and of "**firms competencies**". Whatever the nature of an innovation, its introduction by a firm may be interpreted as a trial (pro-active attitude) or as a reaction (reactive attitude) towards the selection environment. Furthermore, independent of the type of activity of the innovative firm, to innovate requires the activation of specific competencies. To give an example, it can be referred to François *et al.* (1996) presenting **innovation-related competencies** along three axes[18]:

---

16 This synthetic expression can be detailed as follows. **"Innovation is a process"**, innovation-related knowledge is not *a priori* available, thus innovation constitutes a process, an action which is performed and diffused simultaneously **which is: (i) "non maximising":** due to the bounded rationality hypothesis, routines and heuristic behaviours are substituted to optimisation procedures; (ii) **"interactive":** the notion of interaction allows to suppress the apparent contradiction and to make the link between the "demand pull" and "science push" approaches; (iii) **"cumulative":** innovation appears as a dynamic phenomenon which results partly from cumulative processes like learn effects; (iv) **"specific":** innovation-related knowledge and technology are not "manna from heaven" and innovation is perceived as an problem-solving activity; **and (v) "institutionalised":** the institutional dimension of innovation covers the role played by R&D (*i.e.* the institutionalisation of problem-solving activities) and the influence of innovation selection environments (*cf.* Amendola and Gaffard (1988) and Larue de Tournemine (1991)).

17 As it appears clearly in the frame of the evolutionary approach, the term "environment of the firm" refers to a broader and more complex reality than the neo-classical "market".

18 More precisely, those authors propose to take into account two distinct levels of description of innovation-related competencies: *"Nous proposons deux niveaux de description des compétences pour l'innovation: un niveau 'regroupé' des compétences, que l'on qualifiera de complexes, un niveau 'détaillé' de compétences, dites élémentaires. Chaque compétence complexe est décomposable en un ensemble de compétences plus élémentaires. Par exemple, savoir financer l'innovation est une compétence pour innover que nous situons au premier niveau. Elle se décompose en quatre compétences élémentaires: 'savoir évaluer ex ante les coûts de l'innovation', 'savoir déterminer les modes de financement adéquats pour un projet d'innovation donné', 'savoir convaincre les financeurs potentiels tout en préservant le secret sur ce qui doit être', enfin*

(1)   the capacity to "perform";

(2)   the capacity to "learn"; and

(3)   the capacity to "mobilise" external resources.

Cobbenhagen and Den Hertog (1994) provide a more concrete signification to firms' competencies and performance in trying to answer the following question[19]: "*Successful Innovating Firms - What differentiates the front runners?*". Their investigation focuses on firms' core competencies, defined as: (i) organisational competencies; (ii) technological competencies; and (iii) marketing competencies. The authors explain the success and the lack of success of firms[20] with respect to three elements. Firstly, they establish a link between innovation capacity and activation of core competencies. Secondly, they assume that the activation of core competencies reveals a strong cumulative character (*i.e.* core competencies are mutually reinforcing). Finally, and as a consequence, the authors consider that the most innovative firms are the most efficient ones. A complementary analysis is provided by Capaldo, Corti and Greco (1997) who suggest the existence of "virtuous and vicious circles". According to these "circles" (*cf.* figure 1.5), the increase in competitiveness of firms, or the opposite, the running into situations leading potentially to a "final crisis", are strongly determined through firms' ability to innovate. The point of view developed by Capaldo, Corti and Greco (1997) is particularly interesting since it underlines that the introduction of innovation does not lead *mechanically* to improved performance. On the contrary, the decision to innovate may even strongly jeopardise a firm (potentially until the "*final crisis*") .This corresponds fully to the principles evoked previously by Nelson and Winter (1974) about the role of the selection environment. As one may observe, relations between innovation, innovation-related competencies and firms' evolution are complex. This complexity relies strongly on the knowledge-based nature of innovation phenomena.

'*maîtriser tout au long du processus innovant les coûts de l'innovation*' " (François *et al.*, 1996, p. 3).

19   In analysing a sample encompassing 62 manufacturing and service firms.

20   Assessed mainly on the basis of the market position of the considered firms.

Figure 1.5:   **The virtuous circle and the vicious circle**

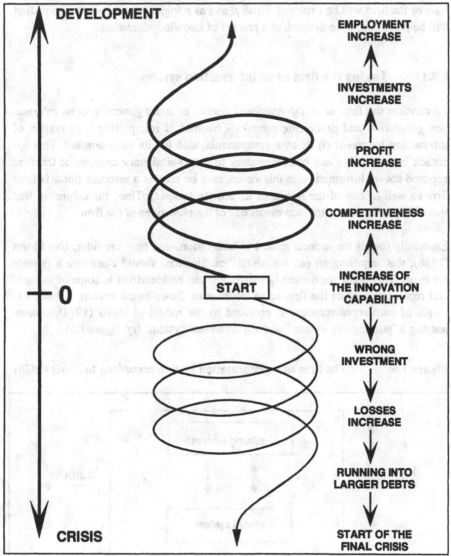

DEVELOPMENT

EMPLOYMENT
INCREASE

INVESTMENTS
INCREASE

PROFIT
INCREASE

COMPETITIVENESS
INCREASE

0

START

INCREASE OF
THE INNOVATION
CAPABILITY

WRONG
INVESTMENT

LOSSES
INCREASE

RUNNING INTO
LARGER DEBTS

CRISIS

START OF THE
FINAL CRISIS

Source: Capaldo, Corti and Greco (1997), p.6

1.2   Innovation as a knowledge-based process

As one may observe, relations between innovation, innovation-related competencies and firms' evolution are highly complex. In order to allow a better understanding of these phenomena, this section will examine innovation - in the lineage of the evolutionary approach - as a knowledge-based process. Consequently, the first section will present the firm as an information system. Then the nature of the knowledge-base of the firm will be explored. Finally, and as a logical consequence, innovation will be presented, or re-defined, as a process of knowledge creation.

1.2.1   Taking the firm as an information system

To consider the firm as an *informational nexus* - or more generally as an *information generation and processing complex* - consists of interpreting it as **system of interaction between: (i) its own components, and (ii) its environment.** This approach is compatible and complementary to the evolutionary analysis of the firm exposed above. Information, in this vision, may be seen as a resource (input) of the firm as well as one of the results of its activity (output). Thus, the control of that specific "production factor" represents one of the objectives of the firm.

Especially for all the aspects related to innovation, one may consider, like Morin (1986), that acquiring an environmental "intelligence" should constitute a priority for firms. This quest for mastering innovation can be identified in some of the several representations of the firm as an information flows-based system. A basic example of such representations is provided by the model of Davis (1974), encompassing a "piloting sub-system" and a "piloted sub-system" (*cf.* figure 1.6).

Figure 1.6:   The firm as an information system according to Davis (1974)

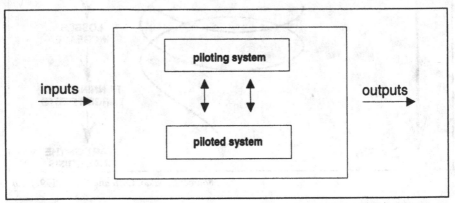

Adapted from: Djellal (1993), p. 46

However, despite the advantage of its simplicity, this representation contains a weakness: the production and memorisation of information by the firm are neglected. Le Moigne (1986) compensates this weakness by integrating the memorisation function of the firm. This additional function allows, from a theoretical point of view, the firm to build its own "decision complexes"[21], initiative and autonomy capacities (*cf.* figure 1.7).

Figure 1.7: **The firm as an information system according to Le Moigne (1986)**

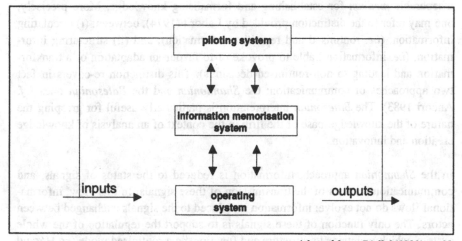

Adapted from: Djellal (1993), p. 46

The introduction of the memorisation capacity leads to consider the firm as an "intelligent information system".[22] This allows the proposal of a definition of the innovative firm which complements the evolutionary approach. In this respect, an **innovative firm** may be characterised as performing two functions: (i) the **control of information** (*i.e.* external and internal information) in order to support the accumulation, application and evolution of its knowledge; and (ii) the **generation of a specific form of "new information"**, (*i.e.* innovations) as the consequence of the evolution of its knowledge-base. In order to validate this definition, the knowledge-base of the firm and the relationships between innovation and knowledge-base will be examined.

---

21 This "decision complexes" recalls the "decision rules sets" evoked by Nelson and Winter (1974).

22 As depicted by Larue de Tournemine (1991). In a similar way, Alter (1990) applies the term "informational enterprise". Additionally, it should be highlighted that this evolution has not only a conceptual character, but also depends on the socio-historical context. Djellal (1993, pp. 48-50) underlines the historical succession, within firms, of three kinds of "logic" related to information during the past fifty years: (i) the *technical logic*, aiming at increasing the volume of information collected and treated; (ii) the *management logic*, improving and standardising information processing modes; and (iii) the *social logic*, corresponding to the era of personal computing and boosting organisational efficiency.

1.2.2   Information and the knowledge base of the firm

Nonaka (1994) underlines the necessity to distinguish clearly between "information" and "knowledge", although they are sometimes used as synonyms. In this respect, information can be considered as a flow of messages or meanings *which might add to, restructure* or *modify* knowledge. Information is thus a necessary and inseparable *medium* for establishing and formalising knowledge. More precisely, one may refer to the distinction provided by Laborit (1974), between: (i) circulating information, (*i.e.* routinised and repetitive information); and (ii) structuring information, (*i.e.* information liable to provoke or to favour an adaptation or a transformation and leading to non-routinised decisions). This distinction re-covers in fact two approaches of communication: the *Shannonian* and the *Batesonian* ones (*cf.* Ancori 1983). The *Batesonian* interpretation is particularly useful for grasping the nature of the knowledge-base of the firm in the context of an analysis of knowledge creation and innovation.

In the *Shannonian* approach, information is reduced to the status of signals, and communication consists of the transmission of these signals. In this view, informational flows do not evolve: information is reduced to the signals exchanged between actors. The only function of these signals is to support the regulation of the whole system, which maintains its structure and functions in a routinised mode (*cf.* Héraud 1987, 1988). The basic information (signals) exchanged are necessarily standardised: the receiver knows in advance all the values or forms that the signals may take. In other words, the universe of possibilities is pre-determined. Therefore, the system operates without "informational surprises", and consequently without innovation. Conversely, the *Batesonian* approach, without denying that each system maintains its structure with the help of a continuous exchange of basic elementary information, integrates the concept of dynamic informational flows (*cf.* Ancori 1983). This means that informational flows, leading to "alterations" of the system are introduced in the analysis. The aptitude of a firm to organise its informational flows and to "extract them from/combine them with" its informational stock can be seen as the expression of its **knowledge-base**.

This leads to two observations which are particularly relevant to the forthcoming investigation. Firstly, the above developed view of information and the related conception of the knowledge-base of the firm authorises the existence of **shared learning effects**. Shared learning effects correspond to situations where learning occurs simultaneously within and outside a given organisation, for instance, when learning takes place in two firms at the same time. It can be quite easily assumed that shared learning effects may influence the innovativeness of firms. The aspects related specifically to shared learning and knowledge exchanges are analysed in the next chapter which is devoted to innovation interactions. The second observation concerns the "nature" of the knowledge encompassed in the knowledge-base of the firm. The generation of new knowledge or more generally its expansion (as shown in figure 1.8) reveals a **non-homogeneity of the knowledge-base of the firm**. Basically, at least a dichotomy can be established, distinguishing: (i) explicit or codified knowledge (*i.e.* transferable and explicable by a code, a scientific or technical language); and (ii) contextual or tacit knowledge (which is the expression of a specific, for instance socio-cultural, context. The implications of this distinction for the innovation activities of the firm are the focus of the next section.

Figure 1.8:   A context of expanding knowledge

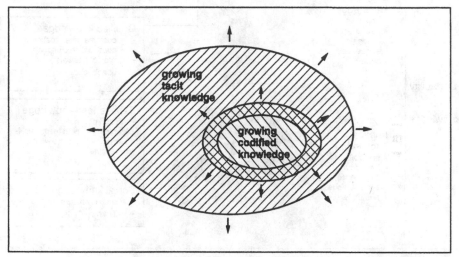

Source: Muller, Zenker and Meyer-Krahmer (1998), p. 6

1.2.3   Knowledge-base of the firm and innovation

"Any organization that dynamically deals with a changing environment ought not only to process information efficiently but also create information and knowledge", asserts Nonaka (1994, p. 14). Knowledge is a necessary pre-condition (i) to understand; and (ii) to create information and is thus intimately interrelated with innovation processes. As expressed above, the knowledge-base of a firm can be seen as a combination of *tacit* (or implicit) and of *codified* (or explicit) knowledge. This distinction (*cf.* the "knowledge pyramid" presented in figure 1.9) has been established by Polanyi (1966). Whereas **codified knowledge** is easily transmittable in a formal and systematic language (comprising words, figures, etc.), **tacit knowledge** has always an implicit or individual related character (strongly based on personal experience) which makes its formalisation and exchange difficult. The knowledge base

of a firm can be expanded by the exploitation of internal search capacities or by the acquisition of external knowledge (*cf.* Saviotti, 1998).[23]

Figure 1.9:   **The knowledge pyramid**

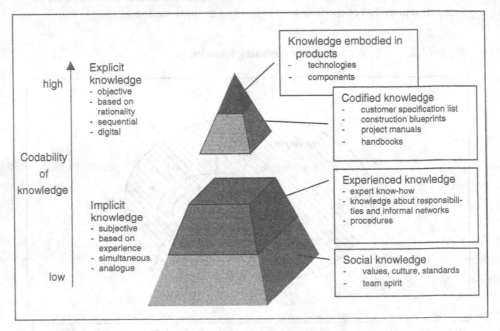

Adapted from: Gassmann (1997, p. 152)

To make the link with the approach adopted by evolutionary economics, it is possible to assert that: (i) firms are organisations which apply different inputs, one of the most relevant for innovation being *information*; (ii) information is *accumulated in* and *processed by* the knowledge base of the firm; and (iii) *knowledge accumulation* and *knowledge processing* by firms results from learning. Additionally, as in the evolutionary approach, cumulative aspects can be underlined. From a dynamic perspective, knowledge can be seen as expanding by associating under different forms, tacit and codified knowledge.[24] On the one hand, the codification of tacit knowledge allows an availability of knowledge which increases with time. On the other hand, the dynamic expansion of codified knowledge generates the apparition of new areas of tacit knowledge.

---

23 In this respect, the expansion of the knowledge-base recalls the principle of "absorptive capacities" of the firm developed Cohen and Levinthal (1989) and previously evoked.

24 For a detailed analysis of knowledge creation, transformation and diffusion within firms see Nonaka (1994).

To sum up, innovation can be conceived as a *key form of organisational knowledge creation*, or as a result of a (cumulative) process in which the firm creates and defines (new) problems and then actively develops (new) knowledge to solve them. Consequently, the definition proposed previously[25] of an innovative firm can be specified as follows. **An innovative firm is:**

(1) **A firm applying and expanding its knowledge base with internal and external information; and (at the same time)**

(2) **A firm generating a specific "new information" (*i.e.* innovations) which is the result of the evolution of its knowledge-base.**

1.3 Nature and forms of innovation: towards a continuum approach

Succeeding the exploration of the informational nature of innovation and the related examination of the knowledge base of the firm, this section deals with the forms or expressions of innovations introduced by firms. The aim is to examine the relevance for the investigation of the distinctions usually employed to "identify" or to "classify" innovations. It will appear that a broader conception of innovation than the traditional one is required.

1.3.1 A good/service continuum?

The existence of a "technical bias" can be easily detected when surveying analyses dealing with innovation and services.[26] This bias leads, for instance, to focus on innovations developed by the manufacturing industry and adopted by services than innovations developed within the service sector. As underlined by Djellal (1993), this "hardware-oriented" vision mainly over-emphasises the adoption of new telecommunication and information processing technologies and notably neglects organisational innovation processes. Even if the apparition and development of "technological artefacts" like, for instance, computer networks, have an undeniable impact on the evolution of the service sector, it does not seem realistic to reduce this evolution to its hardware dimension.

---

25 I.e. an **innovative firm** may be characterised as performing two functions: (i) the **control of information** (*i.e.* external and internal information) in order to support the accumulation, application and evolution of its knowledge; and (ii) the **generation of a specific form of "new information"**, (*i.e.* innovations) as the consequence of the evolution of its knowledge-base (*cf.* section 1.2.1).

26 Mainly: (i) the role of services in the diffusion of innovations and; (ii) the introduction or adoption of innovations by service firms.

In fact, this "technical bias" traduces, on a theoretical level, a restrictive view of innovation phenomena. Gadrey *et al.* (1993) put forward three reasons to explain the deficient knowledge related to innovation and R&D in service firms. Firstly, from an historical perspective, the authors identify the persistence of a view of research based on natural sciences and taking place (exclusively) in the manufacturing industry. Structures which are physically perfectly identifiable like laboratories or R&D centres, are associated with this view. Secondly, Gadrey *et al.* (1993) consider that investigations mostly focus on the production of physical goods. From a micro-economic perspective, this can be interpreted as a view of the firm inherited from the neo-classical approach. Finally, and from a macro-economic perspective, the deficient knowledge concerning innovation in services corresponds to a conception of the economy centred on the "industrial base" (*i.e.* the *manufacturing industry*) which relegates services to the periphery of "productive activities". The traditional conception of goods and services confers the status of a driving activity only to manufacturing. This approach not only sets the production of goods against the production of services: services are considered as subordinate to goods.[27] However, as exposed above, this traditional conception appears quite unsatisfactory and is strongly criticised.[28] Thus, the introduction of the concept of a goods/services continuum may help to go beyond this biased vision of services.

Three main arguments support the renunciation of the principle of the subordination of services, and, at the same time, the recognition of a **"goods/services continuum"**. The first argument deals with the growing inter-penetration of goods and services: for instance, services constitute a major component of the value of most goods. A second argument refers to the role played by information as production resource: information, considered as an input, is required for both the production of goods and of services. The third argument relies on the impossibility to reduce goods and services to a dichotomous scheme opposing "pure goods" (*i.e.* which do not include any form of service) to "pure services " (*i.e.* without any relation to goods). Thus, instead of keeping in mind the "good *vs.* service" stereotype, it seems preferable to adopt a vision in terms of "good/service continuum". The "tangible-intangible dominant continuum" (*cf.* figure 1.10) proposed by Shostack (1977) corresponds to such a vision. The examples proposed provide a convincing illustration that goods and services have dual characteristics combining tangibility and intangibility. This point of view is also supported by Coffey and Bailly (1992, p. 866) who assert that: "*A modern economy must be regarded as an integrated system in which*

---

27   The recognition and adoption of such an approach would have negative consequences for the present investigation of innovative links between SMEs and KIBS, since it would (partly or totally) deny: (i) the possibility that services may support (and even trigger) innovations taking place in manufacturing firms; and (ii) the possibility that service firms themselves may be innovative.

28   For the critics, *cf.* between others Gallouj (1992 and 1994), Djellal (1993), Gadrey *et al.*, (1993).

*the fabrication of goods and the production of services are not viewed as dichoto-
mous functions but, rather, as intersecting zones along a continuum".*

Figure 1.10:   The tangible-intangible dominant continuum

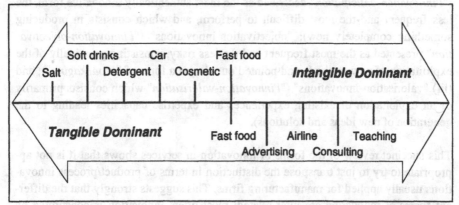

Source: Shostack (1977), p. 77

1.3.2   Innovations of product/services *vs.* process innovations

After having criticised the traditional goods/services dichotomous view, the following step of the reflection questions the validity of the forms most often associated with innovation in the context of the analysis. In other words: is the distinction
between product and process innovation relevant for the analysis of innovation interactions between manufacturing industry and services? Referring to his taxonomy
of innovative (manufacturing) firms, Pavitt (1984, p. 366) asserts: *"[our] proposed
theory also offers an explanation of the balance in different sectors between product
and process innovation".* A clear distinction is made between product and process
innovation, and the "balance" Pavitt (1984) evokes, concerning innovation, is related to the proportion of the different types of firms[29] within the considered manufacturing sectors.

The taxonomy developed by Pavitt (1984) has been transposed by Soete and Miozzo (1990) to services.[30] Nevertheless, the distinction between product (service)

---

29   The four different types of innovative (manufacturing) firms in Pavitt's taxonomy are: (i) sup-
     plier dominated firms; (ii) scale-intensive firms; (iii) science-based firms; and (iv) specialised
     equipment suppliers.

30   The four different types of innovative service firms proposed by Soete and Miozzo (1990, p.
     15) on the basis of the taxonomy developed by Pavitt (1984) are: (i) supplier dominated serv-
     ices; (ii) network services; (iii) production-intensive and scale-intensive services; and (iv) spe-
     cialised technology suppliers and science-based services.

and process innovation loses its relevance. According to Miles *et al.* (1994), if one wants to identify the forms of innovation in services, distinction can be operated between at least: (i) product innovations; (ii) process innovations; and (iii) delivery innovations.[31] In a similar way, Gallouj (1994) proposes a formalisation of innovation activities in the consultancy sector discerning: (i) "anticipation-innovations" ("*l'innovation-anticipation*" described as the most authentically innovative form, the less frequent and the most difficult to perform, and which consists in producing something completely new); "objectivation-innovations" ("*l'innovation-objectivation*" presented as the most frequent and the less risky, consisting principally of the exploitation of new methods and pointed out also as a form of "*repackaging*"); and (iii) "valorisation-innovations" ("*l'innovation-valorisation*" which consists primarily of an exploitation of existing experiences and expertise capacities leading to the generation of new ideas and solutions).

This succinct review of the forms of innovation in services shows that it is not appropriate to try to just transpose the distinction in terms of product/process innovations usually applied for manufacturing firms. This suggests strongly that the different forms of innovation are inter-related: most often, innovation is constituted by the simultaneous existence of several facets. Consequently, the combination of the concept of a good/service continuum with the diversity of inter-related forms of innovation in manufacturing and services industries justifies a broader conception of innovation.

1.3.3   A broader conception of innovation?

According to Miles *et al.* (1994, p. 32) an innovation (*i.e.* a new service or product) requires not only new physical goods "(...) *but requires in addition to technical tools a panoply of tacit operating procedures, understandings and social relations*". In this respect manufacturing and service innovation are similar and, as exposed above, the observation and understanding of the reality of innovation contradicts the traditional conception opposing services and goods.

This leads to renounce on the reduction of innovation to the single introduction of technical artefacts in manufacturing fabrics. Taken as a whole, this approach strongly supports a broader conception of innovation. This conception of innovation by manufacturing and service firms is expressed in the three following points:

---

31 The typology established by Miles *et al.* (1994), specifies that (i) product innovations may derive from process innovations, and most often, correspond to new demand expressed by customers; (ii) process innovations appear as particularly stimulated through new technologies; and (iii) delivery innovations are related to the application of new means and methods as a support to the interaction between the service firm and its customers.

(1) innovation is the result of a process rather than a simple punctual event;

(2) the process leading to innovation depends on the knowledge-base of the firm and consists mainly in information processing;

(3) the innovation process is a constitutive part of the whole evolution process of the firm.

This conception of innovation will constitute a reference guideline for the coming steps of the investigation.

Conclusion

The exploration of the innovation phenomenon had led to underline its specificity as a knowledge-based process. Innovations, as it has been featured, can take place in services as well as in manufacturing firms. Moreover, innovations are likely to affect the evolution of these categories of firms. Resulting from this discussion, the following statements can be considered as a starting point for the empirical investigation: (i) innovation is a complex learning process, based on interactions; (ii) innovation takes place in SMEs as well as in KIBS; and (iii) innovation is an expression of the (positive) evolution of the firm. Consequently, the first main element to be investigated empirically will consist of the search, by SMEs and by KIBS, of a **link between the introduction of innovations and the level of economic performances.**

Chapter 2:   Interactions between KIBS and SMEs and impact on innovation capacities

Introduction

Following the review of the concept of innovation from the point of view of individual firms, the analysis will now turn to interactions between actors. More precisely, this chapter aims at exploring innovation-related interactions taking place between KIBS and SMEs. Firstly, different theoretical approaches to interactions will be confronted: the "make or buy" approach, the transaction costs theory and the network analysis. Then, the impacts of interactions with KIBS and SMEs between each other will be considered respectively.

2.1   Interactions: between market and hierarchy?

The possible choices for firms "preparing" an innovation can be reduced to three options. The first one consists of internal development (of products, processes and more generally of knowledge), the second one of external acquisition (of products, processes and knowledge[32]), and the last one of co-operation or co-development (of products, processes and/or knowledge). The two latter options correspond to situations where actors are interacting. Different analytical settings, i.e.: (i) the "make or buy" approach; (ii) the transaction cost theory; and (iii) the analysis in terms of networks help to better understand the nature of innovation interactions.

2.1.1   The "make or buy" approach

The so-called "make or buy" analysis mainly aims at facilitating the decision process of firms in a situation reduced to the choice between two options. For instance, in a situation dealing with the adoption of new technologies or with the conception of a new product, these two options can be summarised as follows: (i) the option *make* corresponds to internal development; and (ii) the option *buy* means external acquisition.

---

32   The underlying hypothesis here is that knowledge may be easily "acquired", without previous or parallel development of internal knowledge (which in fact depends on the absorptive capacities of the firm, *cf.* Cohen and Levinthal, 1989)

From a general point of view, the "make or buy" analysis is based principally on cost arguments.[33] It is the comparison between acquisition-related costs and internal development-related costs which determines the "make *vs.* buy" decision. However, this vision may generate some criticisms. In fact, while the *a priori* evaluation of acquisition-related costs seems to be possible, it may be more difficult to assess the cost of an internal development. The approach dealing with internal development of new products or processes proposed by Von Hippel (1988) can be considered as an extension of the "*make or buy*" analysis. It consists of a generalisation of the evaluation of innovation-related costs, including an estimation of innovation financial return (or "*expected innovation rents*"). Consequently, the decision of the potential innovator depends on its market position (for instance, monopolistic or quasi-monopolistic situations): the firms introducing innovations will be the ones expecting the highest innovation rents.

Nevertheless, the "make or buy" approach is limited by at least three aspects. The first limit concerns the claimed objectivity of the methods employed. In fact, the results of "make or buy" analyses may vary strongly according to the time horizon chosen, and the precision of costs calculation is subject to strong uncertainties (especially in the case of highly innovative products or services). As a second limit, it is necessary to underline that the application of such methods implies the neglect or even the ignorance of non-quantifiable factors. With reference to the aspects (presented in the first chapter) extracted from evolutionary economics, it appears that non-quantifiable factors[34] strongly determine the success or failure of an innovation. The last main limit of the "make or buy" approach relates to the conceptual absence of a "third option". An additional option (*i.e.* co-operation or co-development) would lead to a more realist range of decision possibilities. Considering these limits, attention has to be paid to further approaches, like the transaction costs theory and the network analysis.

2.1.2   The transaction costs theory

The *Coaseian* conception of the firm goes beyond the "make or buy" approach. According to the views exposed by Coase (1937) in his seminal article "*The Nature of the Firm*", the market system cannot be reduced to an automatic or *self-regulated* system. The neo-classical theory postulates that the co-ordination of agents' decisions is fully realised by market forces. Coase (1937) suggests the opposite in considering that, within firms, market relations tend to disappear due to the co-

---

33   *Cf.* Männel (1981).

34   Like, for instance learning effects in the case of internal development or more generally an expansion of the knowledge-base of the firm in the case of external acquisition.

ordination role of the entrepreneur. This renewed approach to the firm is based mainly on three points ignored by the standard neo-classical theory:

(1) The lack of market transparency, which implies the existence of information research costs.

(2) The necessity for agents to resort to contracts and to ensure that the negotiated contracts are respected.

(3) The additional costs generated by contracts modifications.

Taking these points into consideration, it may be assumed that "being on the market" generates specific costs. In other words, to refer to the market price system implies the apparition of transaction costs. These transaction costs constitute the fundamental reason explaining the existence of firms in the *Coaseian* approach: the substitution of the firm to the market allows the reduction of transaction costs.[35] Nevertheless, the two allocation modes (*i.e.* firms AND market) co-exist which indicates that firms cannot completely replace the market. The *Coaseian* approach explains this situation on the basis of size-costs effects. Organisational costs grow with increasing firm size. However, it is suggested that resource allocation performed only with market transactions are more costly than if they were at least partly performed within a firm. To summarise, an important contribution of the transaction theory is to present the firm not only as a production function but also as an alternative form to the market.

In this respect, Williamson's (1975, 1981) major works can be seen as an extension to the *Coaseian* approach related to the existence and functioning of firms.[36] The *Williamsonian* analysis is based on two key concepts: bounded rationality and individual opportunism. The **bounded rationality**[37] (Simon, 1982) is an attempt to explain why agents have difficulties to find relevant information and to treat it in an optimal way. Due to limitations affecting individuals (for instance in terms of understanding and computation capacities) the "decision-taking rationality" of individuals cannot be perfect in reality. Thus, from a *Williamsonian* point of view, the internalisation of transactions allows the passing from a bounded (individual) rationality to an enlarged (organisational) one.[38] The other key concept, *i.e.* **individ-**

---

35 Considering specifically innovation-related decisions some more precise distinctions may be established, especially between costs related to information search and costs related to contract negotiations. However, the main principles remain the same.

36 For a comparative presentation of the *Coaseian* and *Williamsonian* approaches to the firm, see for instance Koenig (1993, pp. 59-70).

37 The concept of bounded rationality (Simon, 1982) has been previously evoked (*cf.* chapter 1) since this principle already influenced evolutionary views on innovation.

38 It is supposed that enlarged organisational rationality differs from individual bounded rationality, for instance, due to the development of sequential and/or adaptive decision processes.

ual opportunism relies on situations in which an agent does not respect the content of contracts to which he has subscribed (and thus tries to maximise his utility). In the hypothesis of a market encompassing numerous agents, this kind of behaviour may not appear as particularly market-disturbing: the "opportunists" will be excluded from the contract renewal. However, when the market encompasses only few agents (for instance, since the existence of only a finite amount of contractors), individual opportunism leads to asymmetrical transactions, and, consequently reduces market efficiency. Thus, the internalisation of activities (within a firm) appears as a solution avoiding costly negotiation processes related to transactions (on the market).

Thus, the firm constitutes, according to Williamson (1981), a resources allocation mode alternative to the market, in the lineage of the *Coaseian* tradition. As a consequence, the level of transaction costs of a firm depends mainly on:

(1) **The ambiguity of the transaction**: the costs related to the search of partners, to the prevention of opportunist behaviour and to contract supervision increase when needs are specific; when there are only few potential partners and when information is asymmetrically shared.

(2) **The uncertainty inherent in the transaction**: the uncertainty related to the environment in which the transaction takes place affects the costs linked to contract negotiation and redaction.

(3) **The frequency of the transaction**: the existence of learning-effects explains that transactions, if frequent, become less costly.

The transaction cost approach, since it constitutes an analysis in terms of resource allocation and not in terms of resource production, clearly contains certain limitations. According to Herden (1992, pp. 54-62), the transaction cost approach is not sufficient alone to fully understand the nature of innovation interactions. The criticism formulated by Herden (1992) relies firstly on: (i) insufficient empirical evidence and (ii) non-satisfactory reduction of economic reality to the firm/market dichotomy. As a consequence, an additional conception of innovation interactions, the analysis in terms of networks, will be examined.

2.1.3   The network analysis

The development of the concept of networks, and more precisely of **industrial networks**, derives principally from pioneer works like Håkansson, (1987, 1989) who proposes the following definition: "*An industrial network consists of companies linked together by the fact that either they produce or use complementary or competitive products. Consequently the network always contains an element of both co-operation and conflict*" (Håkansson, 1989, p. 16). Håkansson (1987, 1989) distinguishes three constitutive elements of a network. The first element consists of the actors (*i.e.* firms, non-profit-organisations, physical individuals, intermediaries, etc.) which take part in the control, the development, the production, or the commerciali- sation of resources, goods and services within the network. The second constitutive element corresponds to the activities of the network which cover all combination, exchange, transformation or development of resources within the network. The last element is constituted by the resources or network activities inputs.[39] Additionally, the structure of a network can be characterised with the help of four dimensions (Håkansson and Johanson, 1993): (i) the functional interdependency (a network constitutes a system in which actors combine their resources); (ii) the power disper- sion (the power each actor assumes within the network depends on his control of network activities and resources); (iii) the sharing of knowledge (determining the development of network activities); (iv) and the time dimension (a network results from dynamic learning corresponding to an accumulation of knowledge, invest- ments, experiences and routines).

Consequently, a network may be seen as a co-ordination system between market and hierarchy: Within a network, the co-ordination of resources and activities is neither exclusively based on prices (contrary to market relations), nor exclusively on the existence of a hierarchy (contrary to intra-firm relations). Moreover, by con- sidering both direct and indirect relations of an actor within a network, it appears that belonging to a network multiplies the resources to which an actor gets access.[40] The approach in terms of networks insists on the dynamic aspects of co-operation phenomena, whereas, the transaction cost analysis bears mainly on sporadic rela- tions. The specificity of the networks approach consists of the conceptualisation of interdependencies: interdependencies are interpreted as mutual and simultaneous influences of numerous partners in the frame of multilateral relations. One of the motivating reasons explaining the creation of interdependencies between firms con- sists of the diminution (through socialisation) of learning costs. This aspect is par- ticularly relevant for the analysis of innovation interactions: interdependencies may be interpreted as a willingness to concentrate competencies, to collect and select relevant information and thus to support innovation activities.[41] A broader concep-

---

39 Concerning the network activities inputs it is possible to distinguish more precisely between: (i) physical resources (raw materials, equipment, etc.); (ii) human resources (personal, skills, etc.), and (iii) financial resources (capital, investments, etc.).

40 This is, for instance, underlined by Herden (1992, p. 76): "*Über die direkten Austauschbezie- hungen zu den Transaktionspartnern (Kunden, Zulieferer, komplementäre Unternehmen, unter Umständen auch Konkurrenten etc.) entstehen auch indirekte Beziehungen zu verbundenen Unternehmen wie etwa die Zulieferer der Zulieferer, die Kunden der Abnehmer oder die Konkurrenten der Kunden. Diese indirekten Beziehungen erlauben dem Unternehmen den Zu- griff auf - und unter Umständen auch die Kontrolle über - Ressourcen, die außerhalb seines di- rekten Beziehungsnetzes liegen*".

41 This corresponds to the point of view expressed by Julien (1996, p. 5), who asserts that "*Firms cannot innovate if they cut themselves off or do not maintain their innovative contacts, if their*

tion of networks can also be considered[42], making the link with the approach of innovation and knowledge adopted by evolutionary economics (and presented in chapter 1). The "RTE approach" *(réseaux technico-économiques)* proposed by Mustar and Callon (1992) mainly associates three poles: (i) a scientific pole; (ii) a technical pole; and (iii) a market-oriented pole.[43] According to Mustar and Callon (1992), and in the lineage of the concept of *"learning by interacting"* (Lundvall, 1988) it is the collective nature of information which explains the emergence of networks (and indirectly of innovative environments) that stimulate the evolution of firms. The aim of information-sharing within a network is to reduce transaction costs related and knowledge production.[44] **Information sharing within the network constitutes for the interacting firms an intermediate organisational form which completes the basic organisational forms** (*i.e.* market and firms). Thus, interacting firms, possibly limit their transaction costs since they may: (i) decentralise within the network their decision taking and management; (ii) limit irreversibilities by offering the possibility to develop links (or partnerships) within the network ; and (iii) favour learning-effects within the network and thus speed up their innovation processes.

The next two sections (2.2 and 2.3) will focus more precisely on innovation interaction between specific types of firms (*i.e.* SMEs and KIBS). However, before entering the details of those interactions, it seems necessary to concentrate on one particular type of actor. The industrial network approach, as well as in the RTE analysis, evokes networking with research institutes, with non-profit organisations, etc., besides inter-firm relationships. Interactions between firms and such institutions will be designed in the frame of the investigation as **networking with Institutions of Technological Infrastructure (ITI)**. ITI, according to Koschatzky and Héraud (1996), can be defined as entities: (i) with a legal identity (private or public); (ii) located in a specified territory; (iii) having a potential technological impact within this territory; and (iv) whose activities provide the input for research and

---

information sources dry up and are not replenished, if their network decline or if information quality diminishes. And if they do not innovate, they decline and eventually disappear".

42   See DeBresson and Amesse (1991) for a review of the main issues of the network approach.

43   However, the authors emphasize the complexity of the analysis in terms of networks and underline that: "*La description d'un réseau technico-économique ne doit pas se limiter aux seules transactions commerciales, aux seules procédures hiérarchiques, aux seuls accords formels de coopération ou encore à des variables plus sociologiques comme la confiance ou la réputation. Elle doit retracer les trajectoires compliquées qui font circuler entre les protagonistes, des textes, des artefacts techniques, ainsi que des compétences incorporées dans des êtres humains*" (Mustar and Callon, 1992, p. 126).

44   To this extent, such a view completes and extends the views expressed by Nonaka (1994) about knowledge. According to Nonaka (1994), increased circulation - within a firm - between tacit

innovation of firms.[45] The main forms of cognitive interactions between ITI and firms can be summarised under a general form of information networking (*cf.* figure 2.1). More precisely, the activities of an ITI can be interpreted, in knowledge-related terms, as consisting of: (i) collecting and understanding *information* flows; (ii) developing *knowledge* bases; (iii) applying knowledge to problem solving activities, (*i.e.* building specific *competencies*).[46] Thus, networking with ITI may play a determinant role for the innovation activities of firms. This role consists of an extension of the knowledge-base of firms performed additionally to internal "search" and "problem solving" activities, (*cf.* Nelson and Winter, 1974). The external elements provided by ITI to the knowledge-base of firms correspond in fact mainly to "exogenous science and invention" in *schumpeterian* terms or to "research and science" in the chain-linked model (*cf.* section 1.1.1). The aptitude of a firm to transform this external support in innovations (considering the broad conception of innovation proposed at the end of section 1.3) depends on its "absorptive capacities" (*cf.* Cohen and Levinthal, 1989). The integration of networking with ITI in the analysis goes beyond the basic dichotomy opposing firms and the market. The additional knowledge gathered by firms from ITI is neither bought externally on market conditions[47] nor generated internally: it is commonly built up through networking.

---

[45] The original development of the concept of ITI, as well as reflections on the possible measurement of their activities on a territorial level, have been performed in the frame of the Upper Rhine Valley (*i.e. Alsace* and *Baden*) which constitutes the area of empirical investigation of the present analysis (*cf.* Koschatzky and Héraud, 1996). Additionally, Héraud and Muller (1998) provide, for the same geographical area, a complementary analysis focusing on certain categories of French and German ITI.

[46] Examples of those functions can be extracted from Héraud and Muller, (1998, pp. 6-7) for certain categories of ITI (*i.e.* the research labs depending on universities or on public research organisations): "*Co-production of new scientific and technological knowledge (joint R&D project); Execution of external R&D (...); Specific service to the firm (using any professional skill or know-how) (...); Personnel training (transfer of human resources for a limited period) (...); General information (codified knowledge; know-whom and other match-making activities).*"

[47] This, of course, does not mean that everything firms can obtain from (or with) ITI is free of charge.

Figure 2.1:   **Possible activities of Institutions of Technological Infrastructure (ITI)**

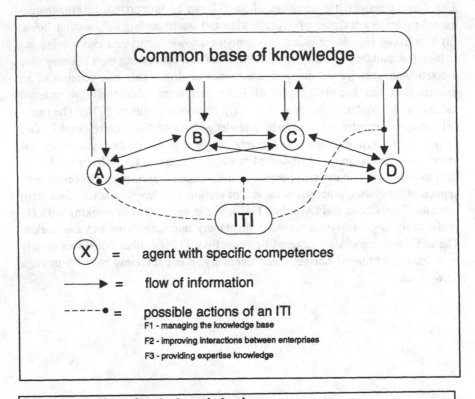

Common base of knowledge

(X) = agent with specific competences

⟶ = flow of information

----• = possible actions of an ITI
F1 - managing the knowledge base
F2 - improving interactions between enterprises
F3 - providing expertise knowledge

Function 1. Managing the knowledge-base
ITI aim at the development of the common knowledge-base of the economic system. Starting from the production of scientific and technological knowledge, this function includes the diffusion of knowledge by educating, collecting and distributing technological information (informing) as well as by guaranteeing the accessibility of the knowledge-base without discrimination (demonstrating).

Function 2. Improving interactions between enterprises
Information and knowledge are not easily marketable commodities. ITI overcome the limitations of the market mechanism in order to adjust supply and demand of technological knowledge and of know-how. In doing so, they can either improve market transactions by using the knowledge of the actors of the system, or provide non-market allocative systems by creating incentive structures. The function of intermediation consists of organising meetings, business fairs, exhibitions and of financing interaction costs in order to improve interactions.

Function 3. Providing expertise knowledge
In this case, ITI are in contact with a single actor, focusing on his very specific needs and providing training or consulting. The aim of training can be to reinforce existing skills or to develop new competencies. Individual support may be related to patenting activities (validating, appropriating) or to providing financial grants (financing).

Adapted from: Koschatzky and Héraud (1996), pp. 3-6

2.2   The impact of interactions with KIBS on SMEs

In this section, the impact of interactions with KIBS on SMEs' innovation activities is discussed. At first, SMEs' knowledge-base and innovation activities are characterised. Then, in a second step, KIBS are considered as complementary innovation assets for SMEs. Finally, the possibility of co-innovations associating SMEs and KIBS is envisaged.

2.2.1   A characterisation of SMEs knowledge-base and innovation activities

SMEs trying to innovate are confronted with several obstacles. In the frame of an empirical study, Kleinknecht (1989, p. 219) provides a list of possible problems which small manufacturing firms might experience in the innovation process. As a result, it appears that the most important limiting factors for SMEs are: (i) capital scarcity; (ii) management qualification[48]; as well as (iii) difficulties to obtain technical information and know-how required for innovation projects.[49] This list of obstacles indicates clearly **that SMEs are confronted with specific limitations in terms of innovation capacity**. Additionally, as Bughin and Jacques (1994, pp. 654-655) underline, failure to innovate is not only related to "bad luck" but seems to be linked to the inability of firms to respect what these authors call "key managerial principles". These "key principles", as stated by Bughin and Jacques (1994), consist of: (i) efficiency of marketing and R&D; (ii) synergies between marketing and R&D; (iii) communication skills, (iv) managerial and organisational excellence; and (v) the protection of the innovation. Consequently, divergences in the abilities of SMEs to activate their innovation potential are related to the way they manage the "key principles".[50]

It is possible to interpret the obstacles to which SMEs are confronted in terms of knowledge-base limitation. Such limitations hinder SMEs in reaching a "critical knowledge-base mass" allowing successful innovation. Thus, their need for infor-

---

[48]   Tested in the inquiry through the statement "costs of an ongoing innovation project are hard to control".

[49]   Additionally, five other barriers to innovation are listed but do not appear as highly statistically significant according to χ^2-tests "(...) 2. *Difficulties in forecasting market demand 3. Expected costs of an innovation project are too high 4. Problems in adapting marketing function* (...) 7. *Problems to find employees with certain qualifications 8. Problems with government regulations*" (*cf.* Kleinknecht, 1989, p. 219).

[50]   This means that innovation activities of SMEs can be seen as subject to "*systematic inefficiencies*, i.e. *firms do not necessarily operate on their best practice frontier*" (Bughin and Jacques, 1994, p. 654).

mational networking, in particular with "knowledge processing agents".[51] The innovation-related information sources on which SMEs rely when applying their knowledge or expanding their knowledge base can be internal or external. Internal sources consist primarily in "search" functions (typically, internal R&D activities) evoked by Nelson and Winter (1974).[52] However, even if internal R&D is efficiently and successfully performed, it is mostly necessary to complete the internal effort with external information or even knowledge. In other words, it is suggested that **internal R&D alone (if any) is not sufficient in most of the cases**. The innovation-related effort, therefore, also consists of the access to external informational resources (*i.e.* primarily the network(s) in which the SME is embedded). This capacity to combine external and internal sources may be interpreted in the meaning given by Cohen and Levinthal (1989), to "absorptive capacities".[53] As an illustration, figure 2.2 presents the scientific and technical information system of SMEs. This presentation, which partly covers aspects evoked previously[54], leads to the examination of the role of KIBS for SMEs' innovation activities. In this respect, in the following, the hypothesis that KIBS constitute complementary innovation assets for SMEs will be developed.

---

51 *Cf.* for instance Julien (1996, p. 2): "(...) *in today's rapidly changing and increasingly global economy, rich information control is becoming a fundamental issue in the survival and development of small businesses. (...) Since most rich information is collective, small businesses can facilitate the process of obtaining and analysing it by joining dynamic information networks. Such networks are particularly effective in innovative environments, where they foster the type of communications that stimulate investment*".

52 *Cf.* section 1.1.3.

53 Concerning the complementary character of internal and external sources, see also Rosenberg (1990) who judiciously asks: "*Why do firms do basic research (with their own money)?*"

54 Notably networking with ITI designed here under "research system" and partly under "industrial system" (*cf.* section 2.1.3).

Figure 2.2:   The scientific and technical information system of a SME

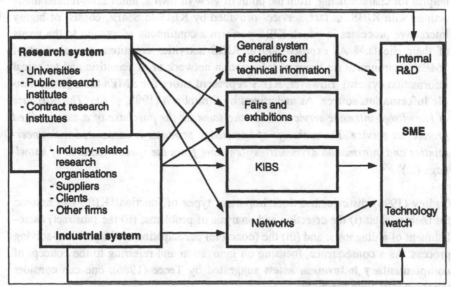

Adapted from OCDE (1993), p. 35

2.2.2   KIBS as complementary innovation assets for SMEs

Adopting an extended conception of innovation (mainly derived from evolutionary economics, *cf.* section 1.3.3), and considering the emergence of a new technological paradigm[55], **interactions with KIBS can be interpreted as a form of complementary innovation assets for SMEs.** Gallouj (1992) highlights that KIBS services should not be reduced to a simple "transfer of technology encompassed in capital"[56] but should be interpreted as an "organisation encompassed in capital".[57]

---

55 This new technological paradigm (in the meaning of Dosi, 1982, extended by Perez, 1983) focuses on *organisational and social innovations*, which are implemented at least partly by services (*cf.* Djellal, 1993, pp. 31-32 and 36-38). This point will be developed further in section 2.3.1.

56 This would correspond to a view in which information is reduced to "stocks" rather than contributing to the evolution of the knowledge-base of the firm as exposed in chapter 1.

57 A similar point of view, focusing on technology transfer, and taking into account services like financial consulting, trade and marketing consulting, patent support, technical support, normalisation and certification support, specific professional training and legal expertise is expressed by Cohendet (1994, p 207) who asserts: *"Il ne s'agit plus de transférer une technologie d'une firme émettrice vers une firme réceptrice, mais de répondre au besoin de développement des firmes, et notamment de développer leurs capacités créatrices et leur faculté d'assimilation et d'apprentissage d'idées nouvelles".*

The concept of complementary innovation assets developed by Teece (1986) is helpful for characterising, from the point of view of SMEs, innovation-related interactions with KIBS. In fact, services provided by KIBS to SMEs, consist of highly interactive processes in which KIBS perform a continuous adaptation to the needs of their clients.[58] As exposed above, KIBS activities constitute one element of SMEs' environment (in terms of information network or of scientific and technical information system). However, **KIBS represent more for SMEs than just a simple information source**. As underlined by Strambach (1998, p. 4): "*The purchase of knowledge-intensive services is not the same as the purchase of a standardized product or service. The exchange of knowledge products is associated with uncertainties and information asymmetries stemming from the special features of knowledge (...)*".[59]

Gadrey (1994) distinguishes three important types of functions KIBS can assume for their clients: (i) the detection and analysis of problems; (ii) the (abstract) establishment of a diagnosis; and (iii) the (concrete) participation to the problem-solving process. As a consequence, focusing on innovation and referring to the concept **of complementary innovation assets** suggested by Teece (1986), one can consider the three following functions:

(1)   KIBS constitute an information source for their clients,

(2)   KIBS assume a "bridge" or interface function between the environment and their clients,

(3)   KIBS reinforce or catalyse evolution and innovation capacity of their clients.

Consequently, it can be assumed that **SMEs interacting with KIBS are in a situation of activation of complementary innovation assets**.

---

58   In this respect, one may refer to Gadrey (1994, p. 33) who declares: "(...) *toutes nos observations concordent pour prouver que lorsque l'enjeu n'est pas la 'réparation' d'un objet ou d'un système matériel, mais un transfert de savoir, tel que l'organisation réceptrice soit en mesure d'utiliser ou d'appliquer les connaissances en question, la réussite de l'opération implique qu'une partie de l'intelligence collective de cette organisation remplisse vis-à-vis des consultants une fonction d'interface*".

59   Additionally, it is important to notice that the interactivity of the "service relation" determines not only the success of the innovation being introduced by the "client", but also allows the development of the knowledge-base of the KIBS. This aspect will be developed in section 2.3 devoted to the impact of interactions with SMEs on KIBS.

2.2.3   KIBS as co-innovators

Going one step further than considering the activation of complementary assets, it can be suggested that KIBS play a role of co-innovators or even of "midwives"[60] for SMEs. This suggestion refers to a broad conception of innovation developed previously (cf. section 1.3.3). Indeed, if innovation activities are not reduced to the sole R&D but encompass other dimensions of the firm (management of human resources, marketing strategy, investment financing, legal protections, etc.), a broader conception of innovation interactions can be considered. This corresponds to the views expressed by models referring to **innovation co-production**.

The *Schumpeter mark III* model proposed by Gallouj (1994) is inspired by the *Schumpeter mark I* and *Schumpeter mark II* models (cf. section 1.1.1). This model attempts to integrate the impact of KIBS[61] on the innovation capacities of their manufacturing clients. The *Schumpeter mark III* model is a **model of co-produced innovation**: innovations performed by the manufacturing firms are (at least partly) developed with the help of KIBS. The *Schumpeter mark III* model encompasses *Schumpeter mark I* and *Schumpeter mark II* rather than replacing them. It offers a conceptualisation of the innovation process (under the forms of two variants, cf. figures 2.3 and 2.4) which is useful for analysing the impact of interactions with KIBS on the innovation capacities of client SMEs.

---

60 Cf. Von Einem and Helmstädter (1994, p. 2) who estimate that: "*Unternehmensbezogene Dienste sind nicht nur selbst ein wachstumsstarker Wirtschaftssektor, sondern sie tragen wesentlich zur Fähigkeit anderer Unternehmen (Industrie und Dienstleistungen) bei, deren Position im Wettbewerb zu stärken, indem sie Informationen, Wissen, Erfahrungen sowie Kapazitäten bereitstellen, die im Strukturwandel gefragt sind. Die wissensintensiven unternehmensbezogenen Beratungsdienste dienen als 'Geburtshelfer', indem sie Rat und Anstöße auf dem Hintergrund allgemeinen Technik- und Managementwissens geben, um es auf den speziellen Einzelfall zuzuschneiden und anzuwenden*".

61 Gallouj (1994) focuses on "consultancy firms". However the spectrum of activities he refers to (*i.e.* legal activities, personal recruitment and management, accounting, strategic consultancy and computer-related consultancy, cf. Gallouj, 1994, pp. 184-188) shows clearly that the conception of KIBS he has in mind corresponds mainly to the one retained for the present investigation.

Figure 2.3:   *Schumpeter mark III* derived from *Schumpeter mark I*

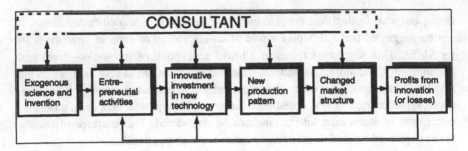

Adapted from: Gallouj (1994), p. 156

Figure 2.4:   *Schumpeter mark III* derived from *Schumpeter mark II*

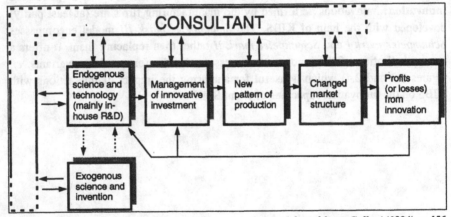

Adapted from: Gallouj (1994), p. 156

On the basis of the *Schumpeter mark III* model it is possible to derive different interaction levels taking place between SMEs and KIBS. The following functions can be fulfilled by a KIBS for a SME:

(1)   The performance of an "environmental or external scanning" related to the innovation activities of the SME. This scanning is not necessarily limited to pure technical aspects, it may also imply other dimensions, like financing, marketing, legal aspects, etc.;

(2)   The performance of an "internal scanning" aiming at a reinforcement of the innovation capacities of the SME;

(3)   The "co-production" of the innovation itself through: (i) contributions in terms of methods applied; (ii) support related to organisational changes and (iii) direct participation in the conception and development of the innovation.

(4)   The "accompaniment" of the SME during the phase of market introduction.

In comparison to the initial *Schumpeterian* conception, this view of co-produced innovations implies three major changes: (i) an evolution in the place of the innovation which moves (partly) from the client firm to the KIBS[62]; (ii) an evolution in the sharing of the risks related to the innovation (a part can be transferred to the KIBS); and (iii) an evolution in the nature of innovation, which is no longer perceived as a purely "technical" phenomenon but also as a development affecting, at least potentially, all the functions of the firm. Trying to provide a synthesis, it can be assumed that the contribution of KIBS to the innovation capacity of SMEs (*cf.* figure 2.5) may be seen as the effect of an interactive service relationship leading to an evolution of client SMEs through: (i) better integration in the innovation environment; (ii) improved activation of internal innovation resources; and (iii) improved activation of external innovation resources. However, the contribution of KIBS to the innovation capacity of SMEs should not be seen as a "mechanical" process. In this respect, one of the main questions to answer is: what turns a SME into "fertile soil" for KIBS contributions? In fact, the existence itself, the intensity and the consequences of KIBS contributions to SMEs depend on several factors. Among other determinants, the role of the structural characteristics of SMEs, the constituents of their knowledge-base and the environmental context of innovation interaction with KIBS will be explored empirically.[63]

Figure 2.5:   The contribution of KIBS to the innovation capacity of SMEs

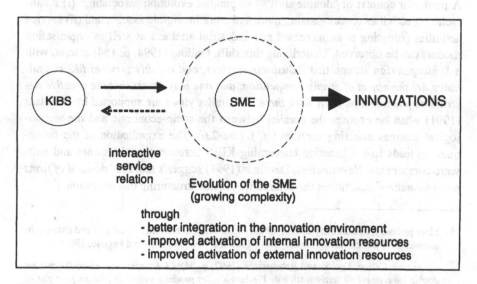

---

62   This shift can be interpreted as an externalisation of the *Schumpeterian* entrepreneurial function.

63   The methodology of the empirical investigation is presented in chapter 4, the results related to SMEs are exposed in chapter 5 and their implications are discussed in chapter 7.

2.3   KIBS innovation capacity and the influence of interactions with SMEs

After having discussed the impact of KIBS on SMEs, focus will be given to the influence exerted on KIBS by interactions with SMEs. At first, the current context of socio-economic and technological changes affecting KIBS' activities will be considered. Then, the knowledge-base of KIBS and its constituents will be examined. As a consequence, the forms of the impact of interactions with SMEs on KIBS will be envisaged.

2.3.1   A context of socio-economic and technological changes

It can be assumed that the current context of socio-economic and technological changes[64] influences particularly service firms (*cf.* Perez, 1983). As a consequence, it can be supposed that innovation behaviours of KIBS (and more generally their evolution) are affected. According to Coffey and Bailly (1992, p. 866): "*The rise of flexible production methods in both goods-producing and service-producing sectors has stimulated the growth of producer service activities; on the other hand, however, increases in the number and variety of available producer services have clearly contributed to the development of flexible production systems.*"

A particular context of "double shift"[65] or parallel evolution associating: (i) manufacturing activities (encompassing more and more intangible assets); and (ii) service activities (adopting at an increased rate industrial artefacts as well as organisation modes) can be observed. Underlining this shift, Gallouj (1994, p. 154) asserts, with a *Schumpeterian* accent, that: "*Business services, and notably services like consultancy are the object of strong competition, and one may assert that the creative destruction phenomenon is on work there*".[66] Similar views are supported by Davelaar (1991) when he exposes the parallel between the socio-economic and the technological changes affecting services (*cf.* figure 2.6). The examination of the consequences leads to a distinction concerning KIBS between their hardware and software components. Nevertheless, Davelaar (1991) suggests that the "*demand of firms for information*" constitutes the determining factor structuring this evolution.

---

64   More precisely, evolutionary authors would evoke shifts in scientific paradigms and changes in technological trajectories (*cf.* notably Dosi, 1982 and Dosi, Marengo and Fagiolo, 1996).

65   *Cf.* as well Callon, Larédo and Rabeharisoa (1997, p. 34): "*L'économie est emportée par un double mouvement de tertiarisation de l'industrie - les produits valent de plus en plus par la qualité et la richesse des services qu'ils fournissent - et d'industrialisation des services - leur explosion est portée par les nouvelles technologies, au premier rang desquelles l'informatique*".

66   "*Les services aux entreprises, et notamment les services de type conseil, sont le lieu d'une importante dynamique de concurrence, et l'on peut dire que le phénomène de destruction créatrice y est à l'œuvre*" (Gallouj, 1994, p. 154).

Figure 2.6:   The place of KIBS in a perspective of socio-economic and technological changes

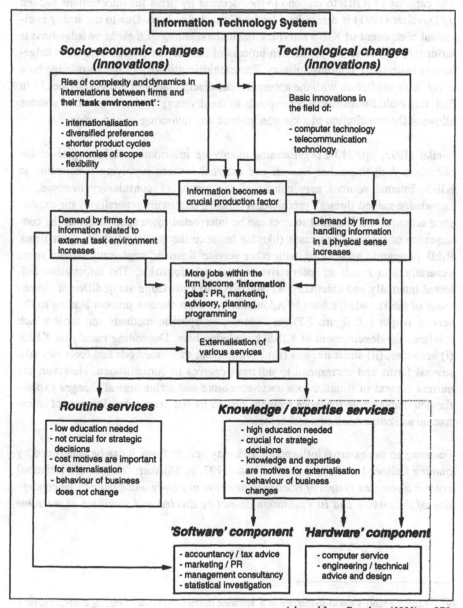

Adapted from Davelaar (1991), p. 279

2.3.2   Exploring the knowledge-base of KIBS

The capacity of KIBS to respond to the "demand by firms for information" evoked by Davelaar (1991) is determined by their knowledge-base. Due to the strong intellectual component of KIBS activities, the understanding of their knowledge-base is intimately related to the production process of the service output. This knowledge-base is necessarily multidisciplinary. The multidisciplinarity of the knowledge-base is not in contradiction with the apparent monodisciplinarity of KIBS activities.[67] In fact, this multidisciplinarity corresponds to the diversity of the information sources allowing the constitution of a knowledge-base and favouring its expansion.

Djellal (1993, pp. 213-214), focusing mainly on information technologies-related consultancy, distinguishes internal and external sources supplying information to KIBS. Internal sources may consist of the results of consultancy missions, of knowledge gained through professional training and more generally of the experience accumulated. External sources can be interpreted more specifically as the consequence of learning processes (like for instance the participation in client firms' R&D programs, networking with other service firms or organisations) and more generally as a **result of interactive service relationships**. The information collected internally and externally is converted into knowledge along different dimensions of the knowledge-base.[68] Additionally, the production process leading to the service output (*cf.* figure 2.7) may encompass specific methods and tools which reinforce the development of KIBS' knowledge-base. Depending mainly on KIBS: (i) activities; (ii) structure; and (iii) environment, these methods and tools can take several forms and correspond to different degrees of formalisation. However, the current context of simultaneous socio-economic and technological changes explain the determining role for those methods played by the drastic development of information and communication technologies.

Focusing on the external influences KIBS may benefit from, it can be underlined by quoting Callon, Larédo and Rabeharisoa (1997, p. 34) that: "*the service* [relation] *creates a complex system of relationships between supply and demand: the conception of the service and its realisation cannot be divided and mobilises at the same*

---

67   KIBS are mostly mono-specialised (e.g. software development, marketing, legal advising, etc.). The (few) firms proposing a wide spectrum of services (for instance from strategic management to computer network installation) constitute an exception. Nevertheless, this exception can only confirm the necessity of a multidisciplinary knowledge-base.

68   This process can for instance lead to the development of competencies, like the ability to establish diagnostics leading to strategic choices or the capacity to provide a relevant organisation or co-ordination of resources to clients.

time the producer and the consumer who co-operate closely".[69] Consequently, the following can be suggested: the experience gained by KIBS from their interactions with their clients (and particularly with SMEs) is one of the elements leading: (i) to the generation of new service realisations, and (ii) to the introduction of modifications in the organisation of the service relation.

Figure 2.7:   The production process leading to the service output

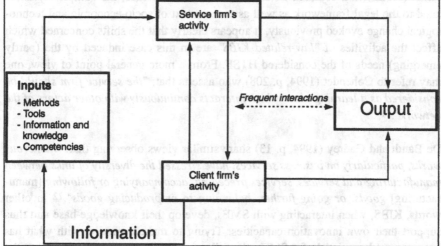

Adapted from: Djellal (1993), p. 214

2.3.3   The impacts of interactions with SMEs on KIBS

In a survey[70] dealing with the determinants of R&D co-operations implying services firms, König, Kukuk and Licht (1996) try to demonstrate that, contrary to manufacturing firms, service firms seem to prefer horizontal to vertical "knowledge flows" (*"Wissensflüsse"*). Results of the survey show that, for instance, in the case of banks and insurance companies co-operations related to innovation activities take mainly place between firms of the same sector (*i.e.* between competitors). Nevertheless, and this exception should be particularly noticed, the surveyed KIBS (*i.e.* software development and technical consultancy firms) did not follow that pattern.

---

69   *"Le service (...) crée un système complexe de relations entre offre et demande: la conception de la prestation et sa réalisation ne peuvent être séparées et mobilisent à la fois le producteur et le consommateur qui coopèrent étroitement"* (Callon, Larédo and Rabeharisoa, 1997, p. 34).

70   The investigation covered a large variety of service firms (*i.e.* distribution, transport, bank, finance, insurance, etc.) as well as two categories of KIBS: software development and technical consultancy.

For the KIBS concerned, the most important R&D partners consist of their customers (for instance manufacturing firms) and not of their competitors. This pleads for the existence of a strong distinctive feature opposing KIBS to most of the service firms.

A particular example of the way relationships linking KIBS to their client may influence those service firms is evoked by Tunney (1998, pp. 33-34). The author, in stressing the "*need for a strategic awareness of European Union Law*" underlines the important role legal consultancy may have for High Tech Small (manufacturing) Firms (HTSF) in a complex environment. In this very case, the complexity is related to the legal framework as well as to a context of socio-economic and technological change evoked previously. It appears clearly that the shifts concerned which affect the activities of "*law-related KIBS*" are in this case induced by the (partly emerging) needs of the considered HTSF. From a more general point of view, one may refer to Cohendet (1994, p. 208) who asserts that: "*the service firm should be considered as a learning place which interacts continuously with other actors of the network*".[71]

De Bandt and Gadrey (1994, p. 13) share similar views observing that: "*numerous works, particularly on business services, have stressed the diversity of links between manufacturing and services, services preceding, accompanying or following* [manufacturing] *goods, or going further, belonging to or producing goods*".[72] In other words, KIBS, when interacting with SMEs, develop their knowledge-base and thus support their own innovation capacities. Trying to make a parallel with what has been assumed previously for SMEs, the following can be suggested concerning the innovation capacity of KIBS (*cf.* figure 2.8). Interactions with SMEs contribute to the evolution of KIBS through: (i) better integration in the innovation environment; (ii) improved activation of internal innovation resources; and (iii) improved activation of external innovation resources. Nevertheless, as has been underlined for SMEs, the contribution of SMEs to the innovation capacity of KIBS cannot be reduced in a "mechanical" way. Several factors may determine this contribution and influence its forms and intensity. The empirical investigation, in examining the impact of KIBS knowledge-base, structure and environment will aim at disclosing the diversity of situations where interactions with SMEs contribute to KIBS innovation capacities .[73]

---

71 "(...) *la société de service doit être considérée comme un lieu d'apprentissage qui interagit sans cesse avec les autres acteurs du réseau*" (Cohendet, 1994, p. 208).

72 "(...) *nombre de travaux, en particulier sur les services aux entreprises, ont fait ressortir la diversité des liens entre industrie et services, les services précédant, accompagnant ou suivant les biens ou, allant encore plus loin, faisant partie de ou produisant des biens*" (De Bandt and Gadrey, 1994, p. 13).

73 Chapter 4 exposes the methodology of the empirical investigation, the results of the KIBS analysis are presented in chapter 6 and their implications are discussed in chapter 7.

Figure 2.8:   **The contribution of SMEs to the innovation capacity of KIBS**

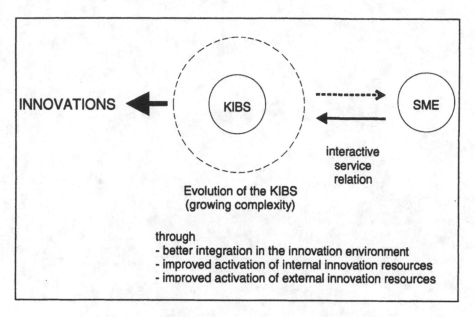

Conclusion

As a consequence of the examination of different and complementary theoretical approaches to the concept of interaction a second constituting element for the empirical investigation is suggested concerning SMEs and KIBS. It is assumed that **interacting SMEs and KIBS have a mutual impact on their respective innovation and evolution capacities**. The relationships linking SMEs and KIBS can be characterised as heterarchical (as opposed to hierarchical). Stressing that heterarchy is one of the conditions allowing mutual learning, Cooke (1998, p. 9) underlines that *"however, heterarchy does not operate in a vacuum"*. This leads to the consideration of the context and environment of such interactions. These aspects, related notably to territorial elements, will be the object of the next chapter.

Chapter 3: Territorial determinants and evolution capacities of SMEs and KIBS

Introduction

In the two previous chapters, innovation has been considered as an evolutionary process as well as an interactive knowledge-based phenomenon. Nevertheless, until this point, the spatial dimension has (deliberately) not been encompassed in the discussion. This will be done in this chapter, which aims at introducing spatial determinants in the analysis of innovation interaction. These spatial determinants are based on two main concepts: (i) the proximity between actors (*i.e.* considering interactions, the closeness according to different dimensions of interacting actors); and (ii) the territory on which actors are located (*i.e.* the elements geographically identifiable as a set which constitutes the reference environment of the considered actors). This chapter examines firstly the concept of proximity with regard to knowledge and innovation. The second section deals with the influence of the territory on innovation activities, considered as a system from a national and regional perspective. The final section exposes a synthesis of the impact of territorial determinants on the evolution of SMEs and KIBS.

3.1   Proximity and innovation

The investigation of the concept of proximity starts with the examination of knowledge spillovers. Knowledge spillovers are considered as an element revealing the incidence of proximity on knowledge- and innovation-related phenomena. This allows, in a second stage, the refining of the question of knowledge accessibility in terms of proximity. As a result, it appears that the coverage of the multiple dimensions of proximity is of crucial importance for assessing its impact on knowledge-related interactions.

3.1.1   Knowledge spillovers and proximity

Knowledge spillovers, whatever forms they take, may be seen as the transformation of organised information, generated, for example, by universities or similar organisations[74], in the form of artefacts, competencies or processes employed by firms. This transformation may particularly be related to innovation processes or even

---

[74] The aspects related to the generation of knowledge by particular institutions has been developed in the section addressing networking and Institutions of Technological Infrastructure (ITI).

correspond to an innovation introduced by the firm benefiting from knowledge spillovers.

The interpretation of knowledge spillovers from a geographical point of view is the object of an intense debate. However, this debate allows a better understanding of some of the dimensions encompassed in the (hypothetical) relations between space and innovation: the proximity between interacting actors. Roughly summarised, if knowledge spillovers appear to be spatially dispersed without a particular pattern, this would suggest that knowledge transmissions are independent from location. If this is not the case, and especially if proximity and density of spillovers are correlated, this indicates that, at least partly, "space matters!".

In fact, the debate on this topic is not over. For instance, on the one hand, Jaffe (1989, p. 968) assesses that: "*there is only weak evidence that spillovers are facilitated by geographic coincidence of universities and research labs within the state*". On the other hand, Anselin, Varga and Acs (1997), by establishing an empirical demonstration based on geographic spillovers taking the form of concentric rings, claim they have proved that "(...) *the positive and significant relationship between university research and innovative activity, both directly, as well as indirectly through its impact on private sector R&D. (...) the spillovers of university research on innovation extended over a range of 75 miles from the innovating MSA[75], and over a range of 50 miles with respect to private R&D*" (Anselin, Varga and Acs, 1997, p. 11). The aim of this presentation is not to survey the literature devoted to the conceptualisation and empirical analysis of knowledge spillovers, but to establish that there is no clear certitude on this subject. As for example Krugman (1991) stressed it: "*Knowledge flows (...) are invisible; they leave no paper trail by which they may be measured and tracked, and there is nothing to prevent the theorist from assuming anything about them that he likes*" (Krugman, 1991, pp. 53-54).[76] This incertitude affects *de facto* the understanding of interactions between KIBS and SMEs.

3.1.2   Proximity, accessibility of information and learning

In fact, mainly deriving directly or indirectly from Isard (1956), a research tradition has been established on the question of economics of proximity and location. Eliasson (1996) provides an illustration of the convergence of these questions with analysis related to the accessibility of information and to learning effects. The ar-

---

75   MSA: Metropolitan Statistical Area.

76   As an answer, and focusing on knowledge codified in the form of patents, certain authors pretend on the contrary that knowledge flows do sometimes leave a "paper trail" (*cf.* Jaffe, Trajtenberg and Henderson, 1993).

guments exposed by Eliasson (1996, pp. 4-5) can be summarised as follows: (i) the locally available but not easily transferable knowledge to exploit business opportunities is a limiting factor for local success and growth; whereas, (ii) communication with competitors (seen as an "active learning mechanism") may lead to an enlargement of the knowledge base. Thus, diffusion and accessibility of knowledge as well as networking and communication procedures play a crucial role for knowledge management. In this context, know-how is distributed by the competitors through the process of "technological competition".[77]

Moreover, according to Antonelli (1998, p. 191): *"Localized technological innovations are the result of differing combinations of tacit and generic knowledge. (...) In fact, many small firms generate significant innovations based solely on tacit localized knowledge; and many larger firms actually fail in the diffusion of innovative initiatives in unrelated activities because of a lack of tacit-learning appropriation opportunities. There is thus a basic need for innovation systems that encourage the accumulation of such tacit localized knowledge and enable its interaction with generic counterparts: in the generation of new technological innovations, firms rely on interactions between themselves and with academic and other research institutions, sharing learning opportunities and experience."*

To summarise, it can be assumed that learning effects, if based on interactions, are not indifferent to the proximity underlying those interactions. Considering the interactive and cumulative character of innovation (*cf.* chapter 1) and the learning dimension of interaction (*cf.* chapter 2) it can be suggested that proximity constitutes an incentive to interact and thus to innovate. However, proximity between actors is not obligatory (only) referring to geographical localisation: diverse meanings (and realities) can be covered. Thus, it is necessary to further explore the concept of proximity itself.

3.1.3 Conceptualisation of proximity

The concept of proximity can be formalised in several ways. A synthesis is exposed for the present investigation on the basis of two different approaches. The first one consists of the distance decay function developed by Staudacher (1991) which fits particularly to SMEs/KIBS interactions (*cf.* figure 3.1). In this model, the impact of

---

77 Complementary elements can be found in McKelvey (1998, p. 165): "(...) *individual firm search and learning is at one level influenced by each firm's relationships and by collective aspects of knowledge. This may be shared with a number of interested parties, such as universities, other firms in the sector, users, etc. However, at another level, the tacit and/or practising aspect of knowledge makes the decision of each and every firm important. A reasonable assumption is that some firms learn faster and more accurately than others, but moreover, firms face different opportunities to innovate in different industries and technologies."*

proximity/distance on the intensity of interactions is a function of: (i) the need of interaction; (ii) the obstacles to interaction; and (iii) the distance between interacting actors. Nevertheless, even if this approach attempts to integrate the different factors modulating its impact, proximity itself remains one-dimensional and thus the model can be characterised as "(log)linear based".

Figure 3.1:   The distance decay function

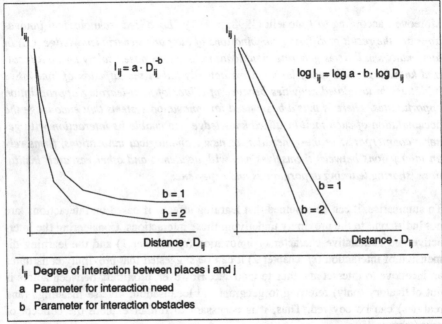

adapted from: Staudacher (1991, p. 144)

The second approach considers proximity in a "multidimensional" perspective. In this respect, the critical approach to proximity, proposed by Rallet (1993) and focusing on innovation processes, can be adopted. The main argument developed by Rallet (1993) is that the role of geographical (*i.e.* physical) proximity can only be analysed with regards to innovation in associating two additional forms of proximity/distance: (i) the "organisational distance" (*i.e.* measurable in the intensity and number of links existing between actors participating to the innovation process); (ii) the "time distance" (*i.e.* measurable by the actual time needed by actors to meet each other). As a consequence, these two additional forms of proximity/distance imply a relativisation of the impact of proximity/distance in the frame of innovation processes.[78]

---

78 In fact even in asserting that "*La contrainte de proximité est d'autant moins évidente que l'on envisage les possibilités offertes par les mobilités conjointes des hommes et des informations*"

3.2   Territory and innovation

To tackle the question of the impact of territorial determinants on innovation, the concept of innovation environment will at first be examined. Then various theoretical approaches exploring the relations between territory and innovation will be reviewed.

3.2.1   The innovation environment and the debated relevance of the territory

The innovation environment of a firm may be defined as the sum of all elements surrounding it and influencing its innovation capacity. The "innovation environment" of a firm must be distinguished from the "innovation system(s)"[79] a firm may belong to. However, the two concepts should be complementary since they are partly overlapping. The innovation environment encompasses and goes beyond the sole selection environment (exposed in section 1.1.3) evoked by Nelson and Winter (1974). In fact, referring to the view provided by Camagni (1991), three "spaces" in which firms evolve may be identified[80] (*cf.* figure 3. 2): (i) the "synergy space" or milieu (*i.e.* the environment of the firm[81]); (ii) the "competition space" (*i.e.* the market); and (iii) the "co-operation space" (*i.e.* the networks in which the firm is involved). In a similar way, Julien (1996) stresses the importance of small firms' integration capacity within their environment. This integration favours shared learning effects, which in turn accelerate the development of an entrepreneurial spirit in facilitating for instance risky decision-making: "(...) *the stronger a small firm links with a dynamic environment, the more innovative it will be, and the greater its stimulating effect on the environment, through a double loop effect*" (Julien, 1996, p. 13). More generally, Malecki (1990) surveying regional innovation-promoting and innovation-inhibiting factors[82] asserts that "*our knowledge of the nature of* [the innovation] *environment is still sketchy, but it is apparent that some*

(Rallet, 1993, p. 378), the author considers that the analysis of the influence of proximity on innovation interactions is, at least partially, biased. According to Rallet (1993), since an innovation can be interpreted as a form of knowledge application and (partially) due to the tacit nature of the knowledge required for it, proximity may matter, but rather from a socio-cultural than from a geographical perspective.

[79]   The concept of territiorialised innovation systems will be exposed in section 3.2.3.

[80]   Apart from what Camagni (1991) designates as the "organisation space" (*i.e.* the firm itself).

[81]   This case corresponds to a particular interpretation of the concept of environment by Camagni (1991), *cf.* the next section for the presentation of the milieu approach.

[82]   Malecki (1990) proposes a review of different studies about promoting and inhibiting factors and concludes by noting that: "*the long-term nature of developing a supportive regional environment is less frequently acknowledged, but it is clearly central to the innovative and entrepreneurial process in creative regions*" (Malecki, 1990, p. 139).

regional environments are more stimulating to innovation and entrepreneurship than others" (Malecki, 1990, p. 127).

Figure 3.2:   Networking and the external environment of the firm

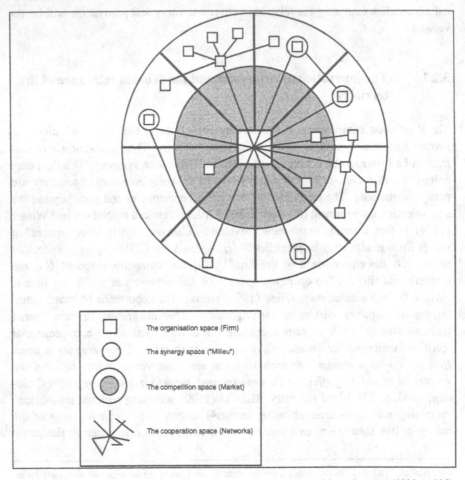

The organisation space (Firm)

The synergy space ("Milieu")

The competition space (Market)

The cooperation space (Networks)

Adapted from Camagni (1991, p. 136)

The difficulty in describing the relations existing between a given territory and the innovation activities of firms located on this territory derives from what Autès (1995, p. 15) calls the *"polysémie du mot territoire"*. In his opinion, the word "territory" covers three distinct (and however related) meanings. Firstly, in a literal meaning, the territory points out the geographical area occupied by humans and their activities. Secondly, the territory has a political meaning, resulting from political (and historical) determinants. In this respect the territory may be governed as an administrative unit identified with the help of borders. Finally, in a symbolical meaning, the territory is populated with "social objects". These "social objects" express: (i) the relationships (of humans, of organisations, etc.) with the two previous forms of territory (i.e. geographical and political); as well as (ii) the way humans and organisations interact with one another.

The multiplicity of the dimensions covered by the concept of territory examines more precisely its signification in terms of innovation-related infrastructure. In the lineage of what has been exposed in section 2.1.3 (devoted to networking) an innovation infrastructure may be defined as follows: it consists of the sum of all institutions which attempt to exert a particular influence on firms in terms of innovation capacities within a given territory. Focusing on the case of specific areas (*i.e.* geographic units lagging behind in terms of development), Capaldo, Corti and Greco (1997) propose a model describing the immaterial infrastructure of a territory (*cf.* figure 3.3). In this approach, each firm located in a specific area can be positioned according to the specificity of its innovation support-related needs.[83] More generally, according to Capaldo, Corti and Greco (1997, pp. 11-12), the innovation infrastructure provides mainly five innovation-oriented functions to firms located on a territory: (i) *"technology transfer"*; (ii) *"creation of new firms"*; (iii) *"education and training services"*; (iv) *"consulting on different company functions"*; and (v) *"sensibilization and promotion in order to transform the latent demand into explicit demand of innovation services"*.

---

[83] The authors use the term "demand of innovation services" for "innovation support-related needs" which do not necessarily correspond to service activities as defined in the frame of the present investigation. On the contrary, to avoid confusions, it is important to underline that in the present **interpretation and utilisation** of the work performed by Capaldo, Corti and Greco (1997), the terms "firm" or "company" may be applied without restriction and thus may designate manufacturing SMEs as well as KIBS. However, it appears that while constituting their model of immaterial infrastructure, the authors had mainly manufacturing firms in mind.

Figure 3.3:   **The model of immaterial infrastructure of a territory**

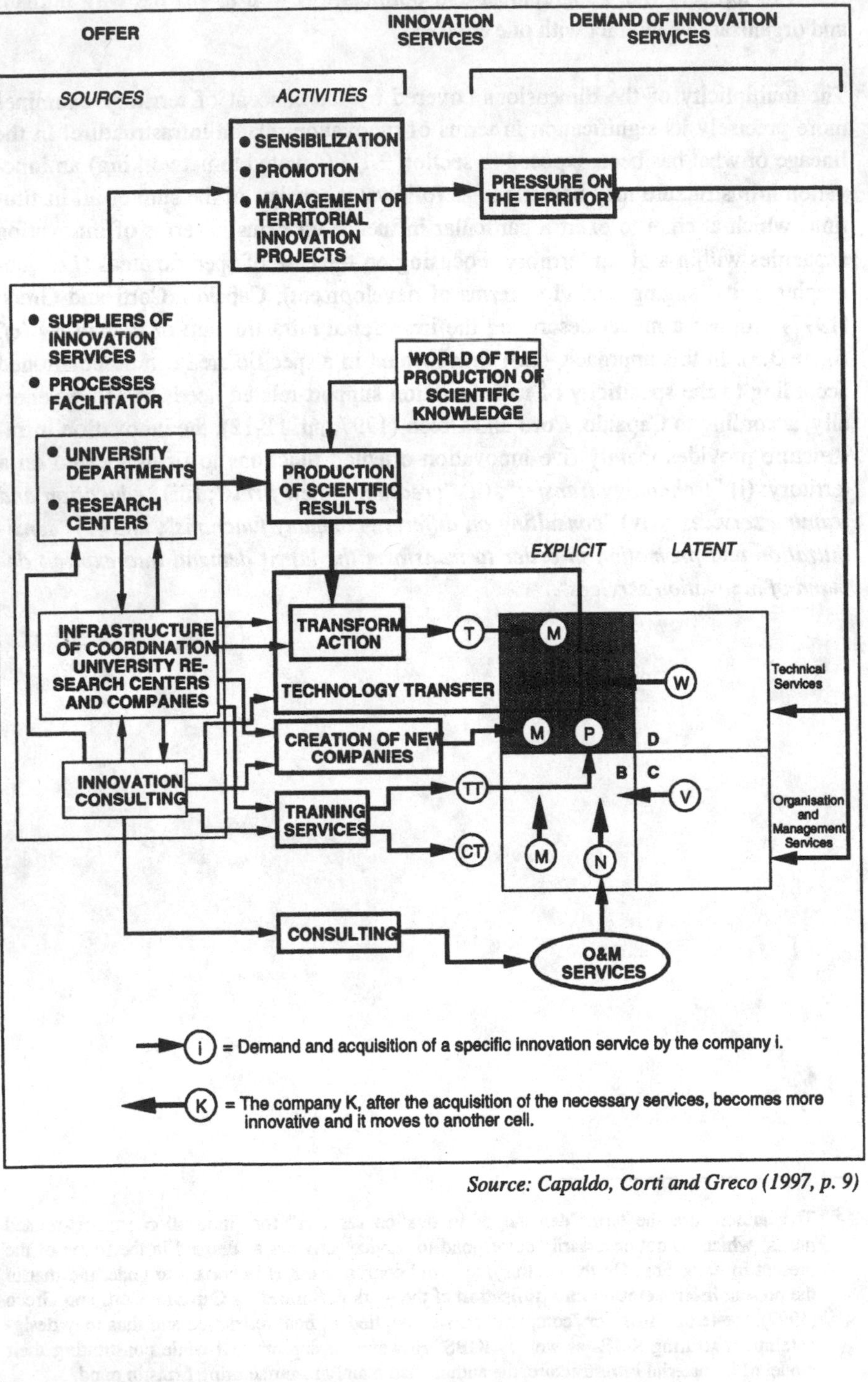

| OFFER | INNOVATION SERVICES | DEMAND OF INNOVATION SERVICES |

SOURCES   *ACTIVITIES*

- SENSIBILIZATION
- PROMOTION
- MANAGEMENT OF TERRITORIAL INNOVATION PROJECTS

PRESSURE ON THE TERRITORY

- SUPPLIERS OF INNOVATION SERVICES
- PROCESSES FACILITATOR

WORLD OF THE PRODUCTION OF SCIENTIFIC KNOWLEDGE

- UNIVERSITY DEPARTMENTS
- RESEARCH CENTERS

PRODUCTION OF SCIENTIFIC RESULTS

EXPLICIT   *LATENT*

INFRASTRUCTURE OF COORDINATION UNIVERSITY RESEARCH CENTERS AND COMPANIES

TRANSFORM ACTION   (T)   (M)

TECHNOLOGY TRANSFER   (W)

Technical Services

INNOVATION CONSULTING

CREATION OF NEW COMPANIES   (M) (P)   D

B C

(TT)   (V)

TRAINING SERVICES   Organisation and Management Services

(CT)   (M)   (N)

CONSULTING

O&M SERVICES

→(i) = Demand and acquisition of a specific innovation service by the company i.

←(K) = The company K, after the acquisition of the necessary services, becomes more innovative and it moves to another cell.

Source: Capaldo, Corti and Greco (1997, p. 9)

3.2.2   From industrial districts to learning regions

Parallel to the innovation infrastructure associated to it, a territory can be characterised with the help of the behaviour and relationships of firms located on it. In the
following, the aim is not to provide an exhaustive survey of the available literature
but to present a selection of theoretical elements illustrating the links between territories and innovation-related behaviour of firms. The apparently old-fashioned
concept of "*industrial district*" (in the meaning of Marshall, 1900) can serve as a
starting point. Benko and Lipietz (1992) observed numerous changes occurring in
the territorial organisation of production recalling strongly *marshallian* industrial
districts. Benko and Lipietz (1992) stress, in particular, two trends explaining the
resurgence of industrial districts: (i) an increased networking of specialised firms;
and (ii) the formation of geographically determined high skilled labour stocks.[84]
This form of industrial organisation[85] characterised by a strong territorial dimension
has been put forward to explain the success of regions like the "Third Italy" (as exposed in Becattini, 1979). In the same way, Piore and Sabel (1984) interpreted the
success of industrial districts as a peculiar case of a general trend, in which flexible
specialisation replaces the Fordist mass-production model, leading to a new spatial
inscription of the organisation of production.

The GREMI[86] approach related to "innovative milieux" (*cf.* for instance Perrin,
1990; Maillat and Perrin, 1992), can be seen as an extension and generalisation of
the notion of industrial district. It corresponds, however, to a great diversity: diversity in terms of systems considered (*i.e.* districts, science-parks, technopoles, etc.)
and diversity in terms of contents attributed to the territory. In the "milieu" approach, the innovative firm does not precede its local milieu but on the contrary, it
is generated by it (Aydalot, 1986). In other words, the milieu is seen as an "innovation incubator". This reformulating of the relations between firms, innovation and
territory in the GREMI approach can be illustrated in three successive stages. The
first stage (Aydalot, 1986; Aydalot and Keeble, 1988) depicts the importance of
local environments and institutional contexts to highlight innovation behaviour. A
further step (GREMI, 1989; Maillat and Perrin, 1992) consists of a specification of
the relations between firms and environment concerning innovation process. The
third stage (GREMI, 1990, Maillat, 1992) deals mainly with the organisational as-

---

84 See also the interpretation of industrial districts according to Becattini (1992, pp. 36-37): "*Le
district industriel est une entité socio-territoriale caractérisée par la présence active d'une
communauté de personnes et d'une population d'entreprises dans un espace géographique et
historique donné. Dans le district, à l'inverse de ce qui se passe dans d'autres types d'environ-
nement, comme par exemple les villes manufacturières, il tend à y avoir osmose parfaite entre
communauté locale et entreprises*".

85 "Industrial" relates on manufacturing activities as well as on services. As BENKO and LIPIETZ
(1992, pp. 15) underline it: "*Un district industriel c'est un district industrieux*".

86 GREMI is an acronym for: "*Groupe de Recherche Européen sur les Milieux Innovateurs*"

pects of collective learning processes leading to innovation. The role of the milieu is reinforced by the conceptual integration of innovation networks in the milieu approach. However, the strong normative character of this view has led to certain critics of the relevance of the milieu approach.[87]

To conclude on territorial aspects and in order to introduce the systemic view, references given here to Lorenzen (1998, p. 26), who makes the link between information-related costs and proximity effects: "*Geographical agglomeration of firms, e.g. located in the same region (and thus speaking same language) and being communities of firms sharing a common socio-economic environment (and thus conception of what has prominence) can be beneficial to sharing of tacit information. The cultural coherence of geographically proximate firms is most likely to be a result of the high degree of interaction between them (...) Business meetings of strategic or formal nature can be held at conference centres or fancy hotels, but discussions of practical nature as well as less formal interactions between agents in co-operating firms may more often happen as daily contact within a certain information contact potential related to geographical distance (...) Thus, the cost of transmitting codified information may have diminished with development of communication and transport technology, but this may not be the case for the cost of communicating tacit information. All in all, geographical proximity may be of crucial importance for information availability.*"

3.2.3   The systemic approach

The systemic approach of the territorial dimension of innovation related activities goes beyond considering the "innovation environment of the firm". Two types of territorial systems will be addressed below: national and regional systems. Systems may be defined in this respect as "*complexes of elements or components, which mutually condition and constrain one another, so that the whole complex works together, with some reasonably clearly defined overall function*" (Edquist, 1997, p. 13). Firstly, the concept of innovation system at a national scale will be addressed, followed by an examination of a complementary view in regional terms.

The nation constitutes *a priori* the most evident type of territory which can be conceptualised (at least in a contemporary western way of thinking). Also, it is not surprising to find in the continuation of Freeman (1987) and Lundvall (1988 & 1992) abundant literature related to national innovation systems (NIS). In the approach

---

87   *Cf.* for instance Rallet (1993, pp. 367-368): "*La problématique* [of innovative milieux] *part d'une définition normative des milieux innovateurs, énonce ce qu'un milieu innovateur doit réunir pour être qualifié comme tel, la proximité territoriale étant le creuset de ses élements constitutifs. (...) Si la démarche éclaire des phénomènes réels indéniables, elle ne rencontre jamais que ce qu'elle postule initialement, à savoir qu'il existe des 'milieux innovateurs'*".

adopted to feature NIS, the basic definition of a system corresponds simultaneously to the elements of the system and the relationships between them. According to this framework, an (national) innovation system consists of elements and relationships that interact in the production, diffusion and organisation of new knowledge.[88] Consequently, an innovation system (whether national or not) is above all a social system within which innovations may be seen as a result of interactions between actors (Lundvall, 1992; Edquist, 1997). However, the literature devoted to NIS does not provide one unique definition of NIS. There are rather several - more or less convergent - conceptualisations related to the ways national contexts affect economic actors notably in terms of innovation behaviour, capacities and success. For instance, Freeman (1987) - taking the example of the Japanese innovation system - particularly focuses on key institutions and on the ways of organising society (*i.e.* education system, government policies, structure of industry and related R&D). According to his views, these institutions and organisations determine in a critical way the introduction of new technologies and the way they benefit the whole (national) system. Another approach explaining why and how nations support innovative activities is proposed by Porter (1990). Four factors appear as conditioning the national system: (i) *factor conditions,* covering a spectrum from the natural resource base to specialised skills and the scientific base; (ii) *demand conditions,* corresponding to the composition and character of demand, seen as potential stimuli for innovations; (iii) *related and supporting industries*, viewed particularly in the meaning of supplier/distributor links which facilitate access to new technologies as well as fostering information flows; and (iv) *firms' strategies, structures and rivalry,* seen as strongly determined by national circumstances. As a synthesis it can be considered that **national innovation systems consist of the sum of all the influences exerted by the firm's overall environment on its behaviour, its interaction patterns and its innovation capacities depending primarily on the national economic, political, social and cultural context.**

Nevertheless, NIS are not the only form of territory addressed by the literature as influencing firms' evolution capacities. "Regional Innovation Systems" (RIS) can be interpreted as a convergence of the concept of NIS and the consideration of a regional level of innovation activities' determinants. Cooke (1998), surveying the literature with focus on the meanings of RIS, underlines that: (i) there is a broad acceptance that classical production systems (e.g. Fordism) are no longer the dominant paradigm of socio-economic co-ordination; and (ii) there is convincing evidence that in a globalised economy, international investment decisions are subject

---

88 *Cf.* for instance McKelvey (1998, p. 172) asserting that: "*A particularly interesting issue here is the question of how much, and when, the individual firm makes decisions which are influenced by collective knowledge and institutions in a national or sectoral system of innovation. (...) each firm's ability to gather and interpret that information differs, or differs in a firm over time, and this seems to partly depend on their (non-market) national environment and on their knowledge relationships with other organizations.*"

to specific regional competitive advantages.[89] These observations lead him to assert that: "*The key conclusion was that, as economic coordination becomes increasingly globalized, the key interactions among firms in specific industry clusters become regionalized.*" (Cooke, 1998, p. 5). However, even if one of the main difficulties related to the analysis of RIS concerns the concept of region itself[90], Cooke *et al.* (1996) assert the following about regions (interpreted as RIS): (i) a region is determined by the size; (ii) a region should be homogenous in terms of specific criteria which distinguish it from bordering areas; and (iv) a region possesses some kind of internal cohesion. The RIS approach encompasses to the greatest extent the concepts of "industrial district", "innovative milieu" and "regional learning". It allows to summarise the arguments in favour of an influence of the regional environment on firms' innovation capacities.[91] RIS can be perceived as a transposition of NIS at the regional level. From both regional and national perspectives, the system of innovation is constituted by elements and relationships interacting in the production, diffusion and use of new knowledge. Consequently, it appears necessary to consider national and regional innovation systems as complementary rather than as opposed.[92]

3.3   The impact of territorial determinants on SMEs and KIBS interactions and evolution

Considering the elements exposed above, three main territorial factors influencing potential innovation interactions between SMEs and KIBS should be examined: (i) the proximity underlying interactions taking place between SMEs and KIBS; (ii) the type of regional environment (characterised in terms of innovation related infrastructures); and (iii) the national innovation system (*cf.* figure 3.4).

---

89   To this point, see also the development by Storper (1997, pp. 169-220).

90   *Cf.* Cooke *et al.* (1996, p. 2): "*Although it has been realised that regional economies are becoming more important, still no general understanding exists of how to define a region*".

91   An alternative perception, which can be found in the literature but not retained here, should be briefly evoked. In fact, it is possible to define a region from an economic perspective, for instance with the help of the approach in terms of clusters (in the meaning of industrial clusters given by Porter, 1990). From this point of view, the industrial cluster may be seen as the sum of all the economic actors contributing directly to the dominant production process of the considered region.

92   Additionally, as Daniels and Bryson (1998, p. 5) put it: "*As production, especially in the advanced economies, has become more flexible there has been a growth in demand for intermediate or producer services that facilitate the organisation and the deepening sophistication of production. They enable and finance trade in goods and in services; they articulate the information and knowledge embodied in the human resources that are central to the new production structures; they are supplied by specialist firms to clients or may be produced internally; and they are pivotal to the spatial organisation of the production system as a whole.*"

Figure 3.4: The mutual contribution of KIBS and SMEs to their innovation capacity: what are the territorial determinants?

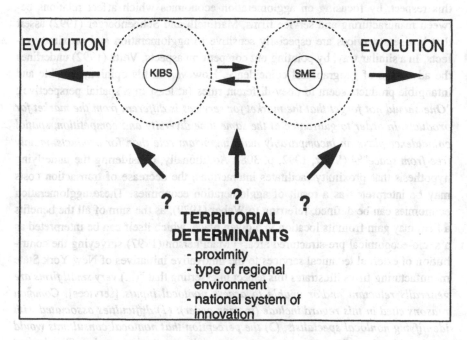

3.3.1 The specific impact of proximity-based interactions between SMEs and KIBS

As seen before, the notion of proximity should be conceived in a "multidimensional" perspective (cf. section 3.1.3). This can be done in considering a territory of reference. A situation in which one SME and one KIBS are both **located in the same territory** should be considered as a situation in which they are "near" or "close". If additionally they are interacting, this can be identified as a situation of **proximity-based interaction**. On the contrary, interacting SMEs and KIBS should be considered as "distant" when they are located in different territories. Consequently, such relationships are designed as non proximity-based interactions. As exposed above, this view is based on the assumption of an intra-territorial homogeneity and an inter-territorial heterogeneity. This assumption concerning homogeneity and heterogeneity relates not only to purely physical distance but encompasses also socio-cultural factors. In other words, and referring to the nature of innovation interactions between SMEs and KIBS (cf. chapter 2), it is assumed that firms located on the same territory have closer behaviour or characteristics and share information and knowledge easier than if they were located on different territories. As a consequence, it is assumed that proximity between KIBS and SMEs constitute an incentive for interaction and implies increased interaction possibilities as well as a reduction of the related transaction costs.

In fact, proximity between SMEs and KIBS supports increased interaction possibilities through easier or more immediate communication and knowledge access. In this respect, by focusing on agglomeration economies which affect relations between manufacturing and service firms, Martinelli and Schoenberger (1992) assert that business services are especially sensitive to agglomeration and proximity effects. In a similar way, by pointing out competition aspects, Veltz (1992) underlines the advantage of geographical coincidence. However, in his opinion, tangible and intangible products seem to obey different rules (at least in a spatial perspective): *"One should not forget that the market for services is different from the market for products. In order to guarantee at the same time diversity and competition, spatial coincidence plays an incomparably more important role than for products, mainly free from space"*[93] (Veltz, 1992, p. 308). Additionally, considering the underlying hypothesis that proximity facilitates interactions, the decrease of transaction costs may be interpreted as a result of agglomeration economies. These agglomeration economies can be defined, referring to Weber (1929), as the sum of all the benefits a firm may gain from its location in a given place, which itself can be interpreted as a socio-economical pre-structured area.[94] MacPherson (1997) surveying the contribution of external technical services to the innovative initiatives of New York State manufacturing firms illustrates this aspect in asserting that *"(...) very small firms are generally reluctant and/or unable to access nonlocal inputs* [services]. *Common reasons cited in this regard include (in rank order): (1) difficulties associated with identifying nonlocal specialists, (2) the perception that nonlocal consultants would not consider these firms important enough to deliver a quality service, (3) a lack of faith in the ability of nonlocal consultants to offer problem-specific inputs, and (4) the perception that nonlocal suppliers are too expensive."* (MacPherson, 1997, p. 66). Besides the benefits proximity may generate in terms of agglomeration economies, it can also affect the nature of the knowledge exchanged between interacting SMEs and KIBS. This may be particularly true in the situations where the concerned knowledge is, at least partly, tacit: *"The ability to assimilate and transfer scientific and technological knowledge that is not completely codified, likewise, is greatly affected by the opportunities for direct personal contact among the parties involved."* (*cf.* David and Foray, 1995, p. 18). More precisely, the specific impact of such proximity-based interactions can be synthesised as consisting of: (i) high opportunities of face-to-face contacts (contrary to "distant relationships"); and (ii) the exploitation of a common "business culture" (expressed under the form of shared

---

93   *"Il ne faut pas oublier (..) que le marché des services est différent de celui des produits (...). Pour garantir à la fois la diversité et la concurrence, la réunion spatiale joue un rôle incomparablement plus important que pour les produits, largement affranchis de l'espace"* (Veltz, 1992, p. 308).

94   To be precise, it must be underlined that the notion of socio-economical pre-structured area implies, however, a broader meaning of proximity than the one adopted here: in this case proximity is not reduced exclusively to geographical coincidence but likely to encompass also cultural and/or behavioural proximity.

values and references, common language, complementary knowledge, etc.). (*cf.* Coffey and Polèse, 1984; Pecqueur, 1989).

It has been indicated previously (*cf.* chapter 2) that interactions between SMEs and KIBS may conduce them to (i) better integration in their respective innovation environment, (ii) improved activation of their internal innovation resources as well as (iii) improved activation of their external innovation resources. Consequently, from the examination of the effects of proximity-based interactions between SMEs and KIBS it can be assumed additionally that proximity between SMEs and KIBS: (i) influences their knowledge exchanges; (ii) may induce a shortage of the transaction costs associated with their innovation-related interactions; and (iii) can be interpreted as an incentive to interact, and thus, through this interaction to innovate for both SMEs and KIBS. To summarise, and referring to those assumptions, the hypothesis can be expressed that, in general, proximity-based interactions between SMEs and KIBS reinforce the contributions to their respective evolution capacities. The empirical analysis will investigate this question and try in addition to identify which characteristics (i.e. in terms of structure and location of firms) influence the sensitivity of SMEs and KIBS to proximity-based innovation interactions.

3.3.2 The influence of the type of regional environment on the evolution capacities of SMEs and KIBS

In addition to proximity aspects, the type of **regional innovation environment** to which firms are exposed may affect their aptitude to innovate as well as their ability to interact (*cf.* section 3.2). This question can be addressed more precisely concerning SMEs and KIBS by considering previous investigations. On the basis of an R&D and innovation survey in the Netherlands (encompassing manufacturing firms as well as service firms[95]) Kleinknecht and Poot (1992) find no evidence that firms located in certain peculiar areas (*i.e.* presented as "*urban agglomerations*") undertake more R&D than firms in other ones (i.e. "*rural areas*"). However, the authors underline that: "(...) *regression analysis shows that service firms in the four major cities are more R&D intensive than service firms in the rest of the country* (...) *the question of whether regions matter for R&D cannot be answered with a simple 'yes' or 'no'.*" (Kleinknecht and Poot, 1992, p. 230). On the other hand, analysing the role of internal and external technical activity in product innovation tendencies[96], Mackun and MacPherson (1997, p. 665) assume that: "*As far as New York State is concerned, empirical evidence from this survey shows that innovation rates are*

---

[95] The empirical basis contained about 4300 firms covering most sectors of manufacturing and service activities.

[96] The valid sample encompassed 396 New York State electrical industrial equipment manufacturers.

significantly stronger in regions that contain rich stock of advanced producer services".

Moreover, in addition to the overall regional environment, the possible influence of the regional infrastructure on KIBS and on SMEs may be direct as well as indirect. The direct influence, potentially affecting all firms within the region and *a fortiori* SMEs and KIBS, consists of an improvement in their evolution capacities (*cf.* sections 2.1.3 and 3.2.1). The indirect or induced influence may be expressed as follows: if SMEs and KIBS benefit from better individual evolution capacities thanks to the regional infrastructure, then this may also reinforce their respective incentives to interact with each other and to benefit from each other (consequently to the hypotheses of mutual benefits advanced in chapter 2). Finally, considering the complexity of the influences of the regional environment, and the difficulties to distinguish causes and consequences, Bellet *et al.* (1992, p. 111) show a resolute certainty in declaring - not without humour – that: *"One may assert, in the most simple terms, that firms' location is not indifferent to their performance, or even more that they derive an advantage from their location on sites where other firms are already established. For which reasons, nobody really knows, but it works!"*.[97]

The contrast between "core" and "peripheral" regions, or more generally **core *vs.* periphery** approaches[98] allow to synthesise the elements gained in the previous discussion. In fact, the idea of a "hierarchy" (or a "ranking"[99]) of regional environments in relation to the evolution possibilities of SMEs and KIBS is of particular relevance for the investigation. From a theoretical point of view, regions, depending on the nature of the innovation system they constitute, influence the innovation, interaction and performance patterns of firms and *a fortiori* of SMEs and KIBS. As a consequence, it can be suggested that territories qualified as "core regions", for instance in reference to their innovation-related institutional and infrastructural dotation, may positively influence the evolution capacities of SMEs and KIBS. *A contrario*, it can be assumed that in regions ranked less favourably[100], SMEs and

---

97   *"On peut affirmer, dans les termes les plus simples, que la localisation des firmes n'est pas indifférente à leurs performances, ou encore qu'elles retirent un avantage de leur localisation sur des sites où se trouvent déjà implantées d'autres entreprises. Pour quelles raisons, on ne sait trop, mais pourtant ça marche!"* (Bellet *et al.*, 1992, p. 111).

98   Numerous "core-periphery models" can be found in the literature; a basic example dealing with endogenous development is provided by Krugman (1991).

99   The principle of "regional ranking" is probably more relevant than "regional hierarchy" for the analysis since hierarchy would imply a dominant/dominated relationship between the concerned territories, whereas the idea of a ranking reflects only specific differences (in terms of advantages and disadvantages) related to the innovation environment of SMEs and KIBS.

100   *I.e.* in regions characterised as intermediate or peripheral according to their innovation-related institutional and infrastructural dotation.

KIBS are less successful in terms of networking, introduction of innovation and economic performance.

3.3.3   The national innovation system as a determinant of SMEs and KIBS behaviour

To consider national innovation systems (NIS, *cf.* section 3.2.3) as a determinant of SMEs and KIBS behaviour refers mainly to specific national characteristics. For instance, a starting point for the cases of France and Germany, is provided by Pilorget (1994). This author, adopting the approach developed by Lundvall (1988) on NIS, proposes a **comparative analysis of France and Germany** concentrating on the ability of the respective systems to support SMEs and to promote innovation-related networks.

This comparative analysis[101] allows to extract particular elements that specifically highlight interactions between SMEs and KIBS. Three main dimensions can be retained: (i) the respective political and economical organisation; (ii) specific technology and innovation-related factors; and (iii) socio-cultural factors. At first, in terms of political and economical organisation, it is clearly possible to compare the German federal structure to the French strongly centralised state organisation reinforced by the economic polarisation towards Paris and the role played by the "*élites*". The second main dimension, (*i.e.* technology and innovation-related factors) relates to the elements developed by Ergas (1986). In this respect, a difference can also be found between France and Germany. In Germany the importance of the technical culture as well as the determinant role of the private sector for R&D activities lead to a system in which technology policies may be characterised as "diffusion oriented" (*i.e.* focusing on the adoption and assimilation of new technologies by firms). On the other hand, in the case of France, due to the determinant role played by the public research programmes and infrastructures (the "*grands programmes*"), technology policies are rather "mission oriented" (*i.e.* focusing on the elaboration and emergence of new technologies). Finally, it is possible to complete the comparison by introducing socio-cultural factors. These factors, beyond history, correspond mainly to differences in linguistic and time/space structures.[102] Schematically, the socio-cultural divergences to be found rely on the importance of creativity in the French case (due to the "polychromic" character of activities), as opposed to

---

[101] The main dimensions constituting this framework in the views adopted by Pilorget (1994, pp. 91-94) for his comparison of France and Germany are: (i) internal modes of organisation of firms; (ii) inter-firms relationships; (iii) R&D systems; (iv) interactions between financial and production systems; (v) role of the public sector; (vi) cultural factors; and (vii) national "ideologies".

[102] This refers to the anthropologic concepts developed by Hall (1959 & 1969) in his seminal works on communication (the "*silent language*") and space (the "*hidden dimension*").

greater precision and efficiency in Germany (explainable by the "monochromic" character of the same activities).

As a result, it can be hypothesised that national differences in terms of national innovation systems lead to distinct interactions and evolution patterns of SMEs and KIBS. More precisely, this would imply for the present investigation that the **German system favours in a stronger way interactions between KIBS and SMEs than the French one.**

Conclusion

Searching to assess the influence of territorial determinants on firms' innovation leads to a conception of a territory based on the "population(s)" located in it rather than on its geographical situation. Nevertheless, the idea of intra-territorial homogeneity has to be confronted with the constraints induced by empirical research. Consequently, concerning the empirical analysis of the evolution capacities of KIBS and SMEs, the following plausible territorial determinants are retained: (i) the proximity/distance between interacting KIBS and SMEs; (ii) the type of regional environment surrounding them; and (iii) the national system to which they belong.

Chapter 4: Operationalisation of the analysis

Introduction

The present chapter is devoted to the operationalisation of the analysis and makes the link between the theoretical reflections and the empirical investigation. Firstly, a conceptual model of the analysis will be developed. This model encompasses the key variables selected for the statistical analysis and is derived from the previously expressed hypotheses. Secondly, the structure of the data collected for the analysis will be examined in terms of regions surveyed and of firm samples. The last section presents the statistical analysis procedure, constituted as a methodology specific to this investigation, combining three distinct data processes successively.

4.1 From the hypotheses to the key variables

In this section, the operational frame of the empirical research is derived from the set of hypotheses expressed in the previous chapters. In a first step, those hypotheses are examined and combined. This leads to the constitution of the conceptual model. Finally, the key variables corresponding to the hypothesis are integrated in the structure of the conceptual model.

4.1.1 The hypotheses to be tested

In chapter 1, it appeared that (i) innovation can be interpreted as a complex learning process, based on interactions, (ii) innovation happens in services as well as in manufacturing firms and (iii) innovating firms should be more "efficient" than non-innovating firms. As a consequence, the first aspect to investigate empirically by SMEs and KIBS concerns the possibility of a **link between the introduction of innovations and the level of economic performance**.

The second main hypothesis derives from the conclusions reached in chapter 2. Due to the specific character (*i.e.* high interactivity and strong knowledge content) of service relations associating KIBS and SMEs, it has been assumed that SMEs and KIBS benefit mutually from those interactions in terms of evolution capacities. For both kinds of firms, this mutual benefit takes the form of: (i) a better integration in the innovation environment; (ii) an improved activation of internal innovation resources; and (iii) an improved activation of external innovation resources. Consequently, it is expected that **interacting KIBS and SMEs show a higher propensity to introduce innovations than non-interacting ones**.

In the light of the debate raised in chapter 3, it can be assumed that the innovation related behaviours and evolution capacities of SMEs and KIBS are affected by determinants of territorial nature. As a result of the discussion, the hypothesis has been formulated that **three main territorial determinants influence the innovation interactions between SMEs and KIBS: (i) the proximity between them; (ii) the type of regional environment surrounding them; and (iii) the national system to which they belong.**

4.1.2   The conceptual model

The combination of the three previous hypotheses in the form of a conceptual model constitutes the core of the empirical analysis. The conceptual model developed to synthesise those hypotheses is firm-based and is assumed to be valid for both SMEs and KIBS (*cf.* figure 4.1). The model is articulated along three main lines which attempt to explain the effects and determinants of innovation interactions between SMEs and KIBS:

(1)   **the structural and environmental determinants** potentially affecting innovation-related behaviours of the firm;

(2)   **the knowledge-base of the firm** depicted with the help of its interaction behaviour the way innovation resources are exploited; and

(3)   **the evolution capacity of the firm** featuring the effects of the combination of behaviour and determinants on its activities and results.

Figure 4.1:   The conceptual model

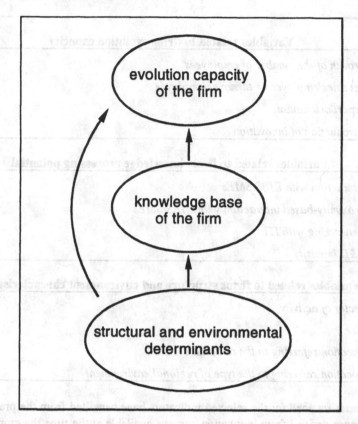

4.1.3   The variables of the analysis

The constitution and the structure of the conceptual model lead to the selection of variables for the empirical investigation. The key variables of the empirical analysis are presented in table 4.1.

Table 4.1:   Overall structure of the variables

Variables related to firms evolution capacity
• *Growth of the number of employees*
• *Net sales/employee or turnover/employee*
• *Export orientation*
• *Introduction of innovation*
Variables related to firms knowledge processing potential
• *Interaction with KIBS/SMEs*
• *Proximity-based interaction with KIBS/SMEs*
• *Networking with ITI*
• *R&D intensity*
Variables related to firms structure and environment characterisation
• *Sector of activity*
• *Size*
• *Location referring to the national system*
• *Location referring to the type of regional environment*

The forms retained for the selected indicators have benefited from the practice and experience derived from innovation surveys available at the time the empirical investigation was constituted (*i.e.* in 1994-1995).[103] However, it is interesting to examine them in the light of indicators and recommendations partly presented afterwards: (i) in Brouver and Kleinknecht (1996) (as a selection of "*Alternative Innovation Indicators and Determinant of Innovation*" in the frame of a reference survey, performed for the Commission of the European Communities, of potential indicators of R&D and innovation in manufacturing and service firms); and (ii) in the OECD manuals devoted to the measurement of scientific and technological activities (respectively the "Frascati manual" devoted to R&D measurement (*cf.* the fifth edition: OECD, 1994) and the "Oslo manual" focusing on business enterprises innovations (*cf.* the second edition: OECD, 1997).

The first set of variables is deals with the **evolution capacity of the firm**. This set associates variables: (i) featuring changes in the firm during the considered time period (introduction of innovations, variation in the number of employees); and (ii)

---

103 For a detailed "survey of innovation surveys" see Le Bas and Torre (1993).

revealing the economic performance of the firm (level of exportation, net sales or turnover per employee). At first, the variable **GROWTH** refers to the growth of the number of employees during the considered three year period. In a similar way Brouver and Kleinknecht (1996, pp. 31-36) consider the variable "growth of a firm's sales", which can also be found in OECD (1997, p. 75). Nevertheless, changes in the number of employees may be seen as less subject to fluctuations than sales and subsequently constitute a worthwhile indicator of the evolution of the firm, particularly when the level of sales per employee is considered. In this respect, the variables **NETSALES** and **TURNOVER** (featuring the level of net sales and the level of turnover by employee respectively for SMEs and KIBS[104]) constitute indicators of firms' performance. The variable **LEVEXP** (featuring the export orientation of a firm, measured the percentage share of exports in total sales) completes the set of firms' performance indicators. It corresponds to the indicator proposed by Brouver and Kleinknecht (1996, pp. 39-40) and to the measurement of sales in terms of domestic and foreign markets recommended by the OECD (1997, p. 75).[105] The variable INNOV relates on the introduction or absence of innovation by each firm during the considered three year period. This is based on own assessment and thus recovers the minimal characterisation of innovation given by OECD (1997, p. 52) *i.e.* at least "new to the firm".[106]

The second set of variables focuses on the **knowledge-base of the firm**. It consists of: (i) the existence of (proximity-based) interactions between SMEs and KIBS and; and (ii) the activation of internal and external innovation-related resources. This distinction between internal and other (external) sources of innovation-related information can be found in OECD (1997, p. 71). The innovation-related interactions between SMEs and KIBS are depicted by the variables **IKIBS** and **ISMES** (for SMEs and KIBS respectively). These variables are completed by **PKIBS** and **PSMES** describing the existence or absence of proximity-based interaction with KIBS and SMEs respectively. SMEs and KIBS are thought to interact on a proximity-base when they are located within the same reference territory (*cf.* section 3.3.1). The variable **NITI** (for networking with ITI, *cf.* section 2.1.3) supplements the observation of the external innovation sources (it corresponds partly to the variable "firm consulted an innovation centre" employed by Brouver and Kleinknecht, 1996, pp. 31-36). Finally, the R&D intensity of a firm referring to the expenses in terms of percentage of its annual turnover (variable **LEVRD**) corresponds to the measure-

---

[104] In this case the constituted variables differ partly between the SME and the KIBS samples: for SMEs intermediary costs such as raw materials and personnel costs have been subtracted in order to minimise the effects of inter-sectoral differences in terms of production methods and thus to provide a more reliable indicator.

[105] More generally, concerning the impact of innovations on the performance of the enterprise *cf.* OECD (1997, pp. 73-75).

[106] At the difference of the distinction (*cf.* Brouver and Kleinknecht, 1996, pp. 22-24) opposing product or service new to the firm and new to the sector of activity of the firm.

ment of R&D expenditures as recommended in OECD, (1994, pp. 91-112) and in OECD (1997, pp. 81-91).[107]

The third group consist of the **structural and environmental variables** assumed to affect the knowledge-base of the firm and its evolution capacities: the structural determinants correspond to firms' characteristics (such as size and sector of activity) while the environmental determinants feature locational characteristics (in terms of region and country). The variable **SECTOR** refers to the NACE classification by firms' main economic activity (*cf.* EUROSTAT, 1993; OECD, 1994, pp. 51-54; and OECD, 1997, pp. 63-65). Based on this classification, 7 and 4 type of activities are delimited for SMEs and KIBS respectively. This approach is similar to the one adopted by Brouver and Kleinknecht (1996, pp. 31-36) who try to integrate sectoral influences in assigning dummy variables to specific firms.[108] The variable **SIZE** follows the recommendations and standards of measurement of the number of employees of a firm (*cf.* OECD, 1994, p. 55; OECD, 1997, p. 66; and Brouver and Kleinknecht, 1996, pp. 31-36). Finally, the location of a firm is described with the help of two variables: (i) **COUNT** (for country) indicates to which national system of innovation a firm belongs (*cf.* section 3.3.3); and (ii) **REGTYP** situates a firm in terms of type of regional environment[109] (*cf.* section 3.3.2).

All the variables are detailed, respectively in section 4.2.2 for the SME sample (*cf.* table 4.5) and in section 4.2.3 for the KIBS sample (*cf.* table 4.9), in terms of precise labels.

4.2   Structure of the data

The data collection was organised in the form of a postal survey, performed in 1995-1996 asking firm representatives to consider activities related to innovation, interaction and general business for the time period 1992-1995. This operation took place in the frame of a broader project entitled "Technological Change and Regional Development in Europe" dealing with regional innovation potentials in the lineage of previous empirical researches devoted to regional networks (*cf.* Herden, 1992; Héraud *et al.*, 1993). This operation was granted by the *Deutsche Forschungsge-*

---

107 An alternative is proposed by Brouver and Kleinknecht (1996, pp. 31-36) in considering R&D man years as a percentage of a firm's total employment.

108 Such as for instance: firm belongs to high opportunity sectors (referring to the typology established by Pavitt, 1984), firm belongs to the service sector, firm's R&D focused on information technology or on biotechnology.

109 *Cf.* Brouver and Kleinknecht (1996, pp. 31-36) who use the dummy variable "location in a central region" in a similar manner.

meinschaft (DFG, the German Research Council) and designed conjointly by three research teams.[110] The overall aim of the research project was to perform a comparative analysis of innovation potentials in three different areas: (i) the so-called research triangle Hannover-Brunswick-Göttingen in Lower Saxony (Germany); (ii) the Federal State of Saxony (in former East-Germany) and (iii) the Upper-Rhine Valley (associating Baden in Germany and Alsace in France). In each of those areas, the research principally took the form of an in-depth investigation of three types of organisations (manufacturing firms, research institutions and producer services) and of the types and intensity of links which could be drawn between them.[111] The current investigation concentrates on a specific part of the data collected: it concerns two regions; Alsace in France and Baden in Germany and two types of firms: SMEs and KIBS.

4.2.1 The surveyed regions

The regions surveyed, Alsace and Baden (*cf.* figure 4.2), both belong to the Upper Rhine Valley.[112] Nevertheless, this common membership to what can be interpreted as a geographical unit in fact covers an important diversity of situations, diversity which is particularly relevant for the present investigation.

It is possible to stress this variety, referring to the argumentation previously developed in chapter 3, with the help of three main elements. Firstly, since the surveyed regions depend on two different countries, two different national systems of innovation are considered. Secondly, the whole area can be subdivided into five sub-regions (corresponding to identifiable territorial units) which can easily be ranked according to the three levels evoked in section 3.3.2.[113] Finally, the geographic reality (in physical terms as well as in reference to the transport infrastructure) of the different sub-regions is strongly compatible with the previously developed ex-

---

110 The joint research project involved the Department of Economic Geography at the University of Hannover, the Faculty of Economics and Business Administration of the Technical University Bergakademie Freiberg and the Fraunhofer Institute for Systems and Innovation Research (FhG ISI) in Karlsruhe. A recent special issue of the review Raumordnung und Raumforschung (RuR 4, 1998) presents different aspects of the methodology and of the results of the overall project.

111 The objectives and theoretical background of the overall project are detailed in Fritsch *et al.* (1998).

112 However, this assertion can be debated in the case of the sub-region of Schwarzwald-Baar-Heuberg. The Upper Rhine "region" or "ensemble" would probably better correspond to reality.

113 The main criteria used to rank the units according to the core/intermediate/periphery scale was the degree of concentration of ITI as well as the consideration of physical infrastructure and administrative and socio-cultural influences. It does not correspond to any form of evaluation of firms or people located in the respective sub-regions.

pression of proximity between two actors, defined as a situation where both actors are located on the same territory.[114]

Figure 4.2:   **The surveyed regions**

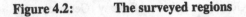

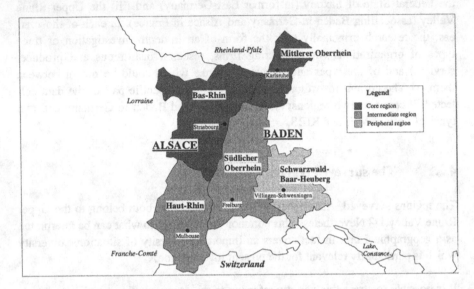

The postal inquiry in Alsace and Baden (in the respective national languages) lead to the constitution of two distinct samples: one encompassing SMEs, the other consisting of KIBS.

4.2.2   The SME sample

The SME sample consists of a total of **n = 726 firms**. All the firms of this sample are comprised of between 10 and 499 employees[115] and are part of the manufacturing sector. Taking the European NACE classification as a reference, it is possible

---

114 Such a situation is not true everywhere, on the contrary, it may even constitute an exception, but it serves the necessity of the empirical test. Altogether, due to the road and rail infrastructure for instance, two firms located in *Haut-Rhin* will be, in most of the cases, "closer" than in a situation where one firm is located in *Haut-Rhin* and the other in *Bas-Rhin* or *Südlicher Oberrhein*. In the same way, interpreting the Rhine river not only as a physical obstacle but also as a socio-administrative and judicial border, it is absolutely realistic to consider that firms, situated only a few kilometers away from Strasbourg but on the German side of the Rhine are "closer" to any other firm in *Südlicher Oberrhein* than to their counterparts in *Bas-Rhin*.

115 It has been decided not to take into account very small manufacturing enterprises (0 to 9 employees). This avoids statistical difficulties related to unreliable data. Experience shows that statistics related to such small firms (supposed to be manufacturing ones) often encompass a large amount of non-manufacturing activities and are therefore strongly biased.

to make the French (NAF) and German (WZ) classifications converge so that the manufacturing industry can be defined as follows (*cf.* table 4.2).

Table 4.2   The activities covered by the SME sample

NACE code (two digits level)	Description
15	Manufacture of food products and beverage
16	Manufacture of tobacco products
17	Manufacture of textiles
18	Manufacture of wearing apparel; dressing and dyeing of fur
19	Tanning and dressing of leather; manufacture of luggage, hand-bags, harness and footwear
20	Manufacture of wood and of product of wood and cork, except fur-niture; manufacture of articles of straw and plaiting materials
21	Manufacture of pulp, paper and paper products
22	Publishing, printing and reproduction of recorded media
23	Manufacture of coke, refined petroleum products and nuclear fuel
24	Manufacture of chemicals and chemical products
25	Manufacture of rubber and plastic products
26	Manufacture of other non-metallic mineral products
27	Manufacture of basic metals
28	Manufacture of fabricated metal products, excepted machinery and equipment
29	Manufacture of machinery and equipment n.e.c.
30	Manufacture of office machinery and computers
31	Manufacture of electrical machinery and apparatus n.e.c.
32	Manufacture of radio, television and communication equipment and apparatus
33	Manufacture of medical, precision and optical instruments, watches and clocks
34	Manufacture of motor vehicles, trailers and semi-trailers
35	Manufacture of other transport equipment
36	Manufacture of furniture; manufacturing n.e.c.

In order to facilitate the analysis[116], the firms of the SME sample have been aggregated according to seven manufacturing categories (*cf.* table 4.3).

---

116 *I.e.* in order to facilitate the interpretation in terms of type of manufacturing activity, but also in order to limit the statistical risks associated with "empty cells" effects.

Table 4.3:   The seven aggregated SME sectors

Aggregated NACE codes	Description
15, 16	Manufacture of food products (and miscellaneous)
17, 18, 19	Manufacture of textiles (and miscellaneous)
20, 21, 22, 36	Manufacture of wood, paper (and miscellaneous)
23, 24, 25, 26	Manufacture of chemicals (and miscellaneous)
27, 28	Manufacture of basic metals (and miscellaneous)
29, 34, 35	Manufacture of machinery and equipment (and miscellaneous)
30, 31, 32, 33	Manufacture of electrical machinery and apparatus (and miscellaneous)

The SME sample (n = 726) is issued from a manufacturing firm sample also including large firms (total n = 766). The overall rate of return of this sample was 17,1 %, which means 16,2 % in France and 17,8 % in Germany. In total in the two regions, 4578 firms were initially identified as manufacturing ones and received the questionnaire, 106 were excluded because they were not relevant and 766 firms (284 in Alsace and 482 in Baden) answered in an exploitable way. It was necessary to exclude 40 firms (17 in Alsace and 23 in Baden) since there size was over 499 employees. This leads to a sample constituted as follows, referring to the sector distribution in each region (*cf.* table 4.4).

Table 4.4.: **The sector distribution of each regional sample (in %)**

Sector	Alsatian population[a]	Alsatian sample	Badian population[b]	Badian sample
Manufacture of food products	21,7	22,4	6,9	3,7
Manufacture of textiles	7,3	6,8	5,9	6,1
Manufacture of wood, paper	17,3	14,4	18,2	16,9
Manufacture of chemicals	14,3	13,3	17,9	12,3
Manufacture of basic metals	16,2	19,4	16,0	22,4
Manufacture of machinery and equipment	13,7	9,9	18,6	19,1
Manufacture of electrical machinery and apparatus	9,4	13,7	16,5	19,5
Total	100	100	100	100

[a] Distribution based on the data provided by INSEE (French statistical office).
[b] Distribution based on the data provided by the IHK (German Chamber of Commerce) of Karlsruhe and Freiburg, detailed data was not available for the sub-region Schwarzwald-Baar-Heuberg.

Examining the sample on the basis of the two criteria of activity and location, no crucial bias affecting its representativity can be detected. However, it should to be underlined that the category "manufacture of food products" is strongly underrepresented in the Badian part of the sample, the category "manufacture of machinery and equipment" is slightly underrepresented in Alsace whereas the category "manufacture of electrical machinery" is over-represented.

The contents of the SME sample in terms of variables and the detailed labels of the concerned variables are shown in table 4.5.

Table 4.5:   Variables extracted from the SMEs survey

Variable code	Variable description	Labels	
GROWTH	Variation in the number of employees (1992-1995)	1- 2- 3-	Increase No variation Reduction
NETSALES	Net sales per employee (1995)	1- 2- 3-	> 35 KECU per capita 25-35 KECU per capita < 25 KECU per capita
LEVEXP	Export orientation (1995)	1- 2- 3-	Strongly export oriented (>25 % of the turnover) Relatively export oriented (5 to 25 %) Not export oriented (less than 5 %)
INNOV	Introduction of innovations (1992-1995)	1- 2-	Introduction of innovation during the period Absence of innovation during the period
IKIBS	Interaction with KIBS	1- 2-	Existence of interactions with KIBS Absence of interactions with KIBS
PKIBS	Proximity with KIBS	1- 2-	Interactions with KIBS mainly proximity-based Interactions with KIBS not mainly proximity-based
NITI	Networking with ITI	1- 2-	Existence of networking with ITI Absence of networking with ITI
LEVRD	Level of R&D expenses	1- 2- 3-	High (R&D expenses > 8% of the turnover) Medium (R&D expenses between 3,5 and 8% of the turnover) Low (R&D expenses < 3,5% of the turnover)
SECTOR	Type of activity	1- 2- 3- 4- 5- 6- 7-	Manufacture of food products (and miscellaneous) Manufacture of textiles (and miscellaneous) Manufacture of wood, paper and printing (and miscellaneous) Manufacture of chemicals (and miscellaneous) Manufacture of basic metals (and miscellaneous) Manufacture of machinery and equipment (and miscellaneous) Manufacture of electrical machinery and apparatus (and miscellaneous)
SIZE	Number of employees (1995)	1- 2- 3- 4- 5-	Less than 20 employees 20 to 49 employees 50 to 99 employees 100 to 199 employees 200 to 499 employees
COUNT	Country of location	1- 2-	France Germany
REGTYP	Type of regional environment	1- 2- 3-	Core region Intermediate region Periphery region

4.2.3   The KIBS sample

The KIBS sample is constituted by **n = 426 firms**. All the firms in this sample employ at least one person. The membership to the knowledge-intensive business services industry is defined for the present investigation referring to the NACE classification (with correspondence to the French NAF and the German WZ, *cf.* table 4.6).

Table 4.6:   **The activities covered by the KIBS sample**

NACE	NAF (France)	WZ (Germany)	Description
72		72	Computer and related activities
721	721Z	721	Hardware consultancy
722	722Z	7220	Software consultancy and supply
723	723Z	7230	Data processing
724	724Z	7240	Data base activities
725	725Z	7250	Maintenance and repair of office, accounting and computing machinery
726		7260	Other computer related activities
74		74	Other business activities
741		741	Legal, accounting, book-keeping and auditing activities; tax consultancy; market research and public opinion polling; business and management consultancy; holdings
7411	741A	7411	Legal activities
7412	741C	7412	Accounting, book-keeping and auditing activities; tax consultancy
7413	741E	7413	Market research and public opinion polling
7414	741G	7414	Business and management consultancy activities
742	742A 742C	7420	Architectural and engineering activities and related technical consultancy
743	743B	7430	Technical testing and analysis
744	744B	7440	Advertising
7484	748J 748K	7484	Other business activities n.e.c.

The originally surveyed population in the two regions included 5776 firms active in the above described sectors. Due to the large number of firms and in order to improve the practicability of the postal inquiry, roughly one third of the firms (*i.e.* 2300 firms) were randomly selected for the survey. The random selection was done according to firms size, location (in national and sub-regional terms) and activity. This selection also gave the opportunity to eliminate dubious data (*e.g.* firms with incomplete addresses or without clear activity identification). With its 426 exploitable fulfilled questionnaires, the KIBS sample corresponds to an overall return rate of 22,1 % (respectively 17,4 % for the French part of the survey and 25,7 % for its German counterpart).

Within the KIBS sample, in order to facilitate the analysis (and for the same statistical reasons leading to the limit of the number of SMEs categories to seven), firms are classified according to four aggregated categories (cf. table 4.7)

Table 4.7:   The four aggregated KIBS sectors

Aggregated NACE codes	Description
721 to 726	Computer related consultancy and activities
7412	Legal, accounting and tax consultancy
7413, 7414	Business, management and marketing consultancy activities
742, 743, 744	Architectural, engineering and technical activities

In order to detect bias affecting its representativity, the KIBS sample is examined on the basis of the two criteria of activity and of location (cf. table 4.8). One main divergence between the observed population and the sample can be established concerning the category "legal, accounting and tax consultancy". A clear underrepresentation in Alsace and an over-representation in Baden appears. These variations tend to compensate themselves when considering the whole sample. However, in the case of statistical analysis focusing on the type of activity or on the national location, this bias may have an influence on the result. Nevertheless, the existence of this (modest) bias, constitutes an expression of some differences between French and German KIBS active as "legal, accounting and tax consultant": it could for instance reveal a greater commitment to innovation or to clients' innovation support of this type of KIBS in the German context in comparison to the French one.

Table 4.8:   The sector distribution of each regional sample (in %)

Sector	Alsatian population[a]	Alsatian sample	Badian population[b]	Badian sample
Computer related consultancy and activities	15,0	16,3	26,7	27,2
Legal, accounting and tax consultancy	34,8	16,3	8,1	16,1
Business, management and marketing consultancy activities	21,5	25,1	31,2	24,7
Architectural, engineering and technical activities	28,6	42,1	33,9	31,9
Total	100	100	100	100

[a] Distribution based on the data provided by INSEE (French statistical office).
[b] Distribution based on the data provided by the IHK (German Chamber of Commerce) of Karlsruhe and Freiburg, detailed data was not available for the sub-region Schwarzwald-Baar-Heuberg.

The contents of the KIBS sample in terms of variables and the detailed labels of the concerned variables are exposed in table 4.9.

Table 4.9: **Variables extracted from the KIBS survey**

Variable code	Variable description	Labels	
GROWTH	Variation in the number of employees (1993-1996)	1- 2- 3-	Increase No variation Reduction
TURNOVER	Turnover per employee (1996)	1- 2- 3-	> 100 KECU 50 to 100 KECU < 50 KECU
LEVEXP	Export orientation (1996)	1- 2- 3-	Strongly export oriented (>25 % of the turnover) Relatively export oriented (5 to 25 %) Not export oriented (less than 5 %)
INNOV	Introduction of innovations (1993-1996)	1- 2-	Introduction of innovation during the period Absence of innovation during the period
ISMES	Interaction with SMEs	1- 2-	Existence of interactions with SMEs Absence of interactions with SMEs
PSMES	Proximity with SMEs	1- 2-	Interactions with SMEs mainly proximity-based Interactions with SMEs not mainly proximity-based
NITI	Networking with ITI	1- 2-	Existence of networking with ITI Absence of networking with ITI
LEVRD	Level of R&D expenses	1- 2- 3-	High (R&D expenses > 8% of the turnover) Medium (R&D expenses between 3,5 and 8% of the turnover) Low (R&D expenses < 3,5% of the turnover)
SECTOR	Type of activity	1- 2- 3- 4-	Computer related consultancy and activities Legal, accounting and tax consultancy Business, management and marketing consultancy activities Architectural, engineering and technical activities
SIZE	Number of employees (1996)	1- 2- 3- 4- 5-	Less than 3 employees 3 or 4 employees 5 to 9 employees 10 to 19 employees 20 and more employees
COUNT	Country of location	1- 2-	France Germany
REGTYP	Type of regional environment	1- 2- 3-	Core region Intermediate region Periphery region

These variables, together with the ones extracted from the SME survey, will be exploited according to the investigation procedure exposed in the next section.

4.3   The statistical exploitation procedure

The **original exploitation procedure** developed for the investigation consists of an *ad hoc* methodology combining three different multivariate statistical exploitation methods (*cf.* figure 4.3). In a first or "explorative" step, selected "segmentation procedures" are performed with the help of CHAID algorithms. CHAID (Chi-squared Automatic Interaction Detector) is an algorithm allowing the analysis of categorised variables based on segmentation modelling. In the second or "intermediate" stage, a multiple correspondence analysis is performed in order to characterise firms' structures, behaviours and results. In the final step, a "path modelling" based on the performance of PROBIT algorithms is proposed.

Figure 4.3:   **The "stat-mix" procedure**

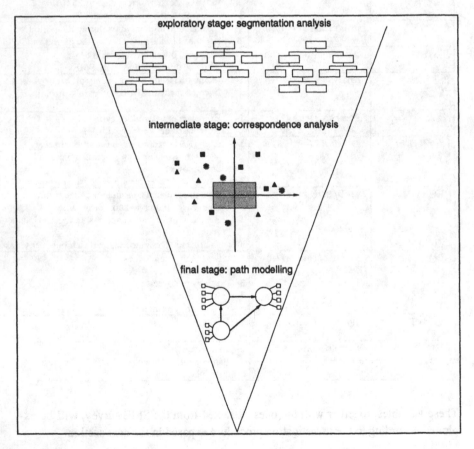

4.3.1 The exploratory stage: the segmentation analysis (CHAID)

CHAID (Chi-squared Automatic Interaction Detector) is an algorithm based on **segmentation modelling** allowing the investigation of categorised variables. The overall goal of this procedure is to detect correlations between different variables with the help of "repetitive and successive" χ^2-tests (such as for instance Pearson's χ^2):

χ^2 **(Pearson):**

$$\chi^2 = \sum_i \sum_j \frac{(n_{ij} - \hat{n}_{ij})^2}{\hat{n}_{ij}}$$

with:
n_{ij} observed frequencies
$\hat{n}_{ij}$ expected frequencies
i, j rows and columns of the crosstabs

CHAID divides the investigated population (sample) into segments that differ with respect to a designated criterion (called the "dependent variable"). The segmentation of the sample into two or more distinct groups is based on the categories of the **best predictor** of the dependent variable. CHAID then splits each of the groups into smaller subgroups based on other predictor variables. This splitting process continues until no more statistically significant predictors can be found.[117] The final subgroups constitute segments which are mutually exclusive and exhaustive. To perform the segmentation procedure, CHAID merges different categories of an independent variable if they are not significantly different. As a result of the merging procedure, the individuals present in a same segment are homogeneous with respect to the segmentation criterion while individuals situated in different segments are heterogeneous relative to the dependent variable. The segments generated by a CHAID procedure can easily be presented in a tree-diagram (*cf.* figure 4.4).

---

[117] See Magidson ; SPSS Inc. (1993) for a detailed presentation.

Figure 4.4:   **Example of CHAID tree-diagram**

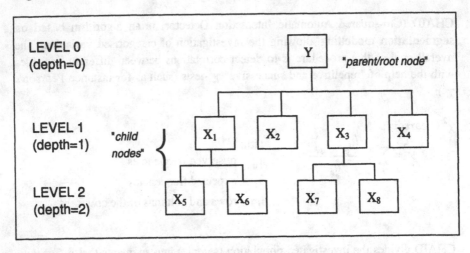

CHAID is particularly useful for detecting interaction effects to be taken into consideration in order to be included for instance in a regression analysis. The interpretation of segmentation results in terms of "causality paths" allows the performance of a selection process within an huge set of variables. This "selective reduction" of variables is particularly relevant as a preliminary step before applying a correspondence analysis.

4.3.2   The middle stage: multiple correspondence analysis

The development of correspondence analysis derives mainly from the pioneer work performed in the 60's by J. P. Benzécri.[118] Originally, such procedures were limited to the analysis of contingency tables (crosstabs of two nominal characters). Meantime, correspondence analysis has been extended to (at least theoretically) an unlimited number of characters. Thanks to their mathematical properties and due to their richness in terms of interpretation potentialities, correspondence analysis constitute a powerful tool for exploiting qualitative data. All variables in a multiple correspondence analysis (also called homogeneity analysis) are inspected for their categorial information only. That is, the only consideration is the fact that some objects are in the same category while others are not. One important advantage (due to the presence of qualitative or categorised variables only) is the possibility of considering non-linear relations between variables.

---

118 See for instance Benzécri (1992) for a detailed presentation and overview of the possibilities in
   this field.

In a multiple correspondence analysis n *elements* characterised through p *(qualitative) characters* are taken into account. To each *character* j, the set of its m_j modalities is associated. This lead to the *disjunctive table* X:

$$X = \begin{matrix} 1 \\ 2 \\ \vdots \\ n \end{matrix} \left(X_1 \Big| X_2 \Big| \ldots \Big| X_j \Big| \ldots \Big| X_p \right)$$

X: *n rows*

$m_1 + m_2 + \ldots + m_p$ *columns*

The vector exploitation of the different sub-spaces constituted and their projection in two dimensions along two axis is based on a generalisation of the canonical correlation analysis. This generalisation leads to the performance of a multiple correspondence analysis of the disjunctive table. This can be interpreted as a correspondence analysis of a particular contingency matrix B (called also Burt's table, where $B={}^tX.X$).

Basically, a canonical analysis consists of considering two sets of characters or vectors of $\Re^n$:

$(x^1, \ldots, x^j, \ldots, x^p)$ and $(y^1, \ldots, y^k, \ldots, y^q)$

and to try to "bring closer":

$\xi = a_1 x^1 + \ldots + a_j x^j + \ldots + a_p x^p$ and $\eta = b_1 y^1 + \ldots + b_j y^k + \ldots + b_p y^q$

in searching for the coefficients:

${}^ta = (a_1, \ldots, a_j, \ldots, a_p)$ and ${}^tb = (b_1, \ldots, b_j, \ldots, b_p)$

so that the square correlation between ξ and η will be maximum.

ξ and $\eta \in \Re^n$ are called *canonical characters*; $a \in \Re^p$ and $b \in \Re^q$ are called *canonical factors*. The set of ξ characters, as a combination of $(x^1, \ldots, x^j, \ldots, x^p)$ constitutes a sub-space vector $W_1 \in \Re^n$ and the set of η characters, as a combination of $(y^1, \ldots, y^k, \ldots, y^q)$ constitutes a sub-space vector $W_2 \in \Re^n$. The research of a maximal correlation corresponds then to the research of $\xi \in W_1$ and $\eta \in W_2$ presenting a minimal angle (*cf.* figure. 4.5):

Figure 4.5:   Canonical analysis, $\xi \in W_1$ and $\eta \in W_2$ presenting a minimal angle

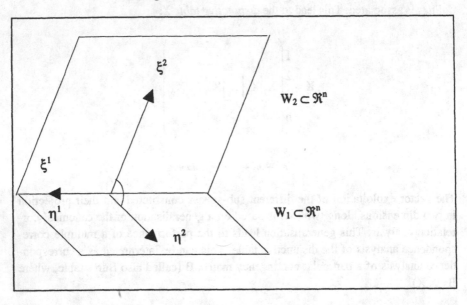

The multiple correspondence analysis is the result of the generalisation (in n dimensions) of the canonical analysis.

4.3.3   The final stage: path-modelling

The path-modelling procedure is based on the successive performance of PROBIT algorithms. PROBIT algorithms estimate maximum-likelihood model in order to detect dependencies associating a dependent variable to explanatory (or independent) variables. The general form of the model to be estimated is:

$$\Pr[E = 1] = \Phi[\beta_0 + \beta_1 A + \beta_2 B + ... + \beta_n X]$$

with Φ being the cumulative normal distribution, E (dichotomic variable) the dependent variable, A, B and X being the explanatory variables. β_0 is the constant, β_1, β_2 and β_n are the coefficients of the independent variables in the equation.

The path-model is obtained by placing selected variables (according to the conceptual model developed) alternatively in the role: (i) of dependent variable in respect to some variables; or (ii) of independent variable in comparison to others. Since the application of PROBIT requires *dichotomic dependent variables* it is necessary to binarise the variables of the analysis. This can be done in a relevant way on the basis of the previous analytical steps, especially in exploiting the results of the multiple correspondence analysis.

The set of "paths" resulting from the procedure can be interpreted as a "picture" of the interrelations between the variables. Variables can at the same time be explanatory (or independent) and dependent, depending on their relative position, (*cf.* figure 4.6).

Figure 4.6:   Path modelling – dependent and explanatory variables

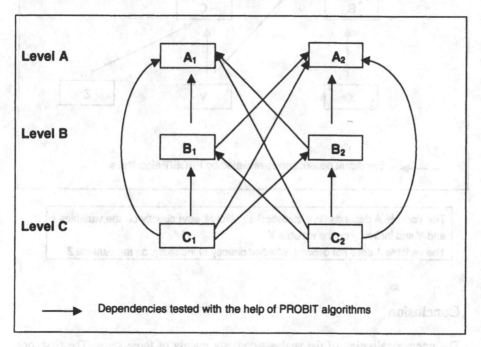

Dependencies tested with the help of PROBIT algorithms

The variables corresponding to level A are dependent variables relative to levels B and C

The variables corresponding to level B are explanatory variables relative to level A and dependent variables relative to level C

The variables corresponding to level C are explanatory variables relative to the levels A and B

As a consequence the identification of significant dependencies allows the interpretation of the revealed "paths" in terms of causality effects, and to distinguish between direct and indirect dependencies (*cf.* figure 4.7).

Figure 4.7:   Path modelling – direct and indirect dependencies

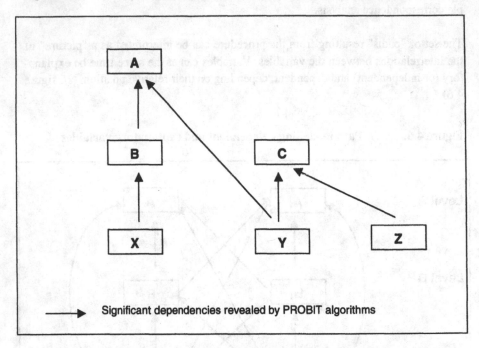

Significant dependencies revealed by PROBIT algorithms

The variable A depends (in a statistical significant way) directly on the variables B
and Y and indirectly on the variable X
The variable A does not depend, whether directly or indirectly, on the variable Z

Conclusion

The operationalisation of the analysis consists mainly of three steps. The first, or
conceptual, step provides a model associating and linking together the hypotheses
derived from the theoretical reflections developed in chapters 1 to 3. The second
step concentrates on the specific data used for the empirical test of the model. The
collection of required information has been allowed by the design of a particular
firms survey. The samples constituted cover populations of SMEs and KIBS spread
over two countries (France and Germany) and corresponding to three different types
of regional environment (*i.e.* core, intermediate and peripheral). Finally, the third
step of the operationalisation is devoted to the data processing methodology. To this
end, an *ad hoc* methodology has been developed, based on the combination of three
complementary statistical exploitation methods (*i.e.* CHAID, multiple correspondence analysis and PROBIT algorithms). Thanks to this combination the results of
each stage of the analysis can be introduced to the next one. Consequently the information gained in using the whole procedure is much richer than just the sum of
the results of each separated data treatment method.

Chapter 5:   Statistical exploitation of the SME sample

Introduction

This first chapter dealing with empirical results is focussed on the SME sample. The methodology adopted is the one described in the previous chapter, combining, in a successive way, segmentation procedures, multiple correspondence analysis and PROBIT algorithms.

5.1   Segmentation procedures

The first step of the empirical investigation dealing with the SME sample is the application of **segmentation algorithms** (CHAID). The aim of this first stage is to allow an exploration of existing correlations between variables. In this way an indication for the process of selection of variables is gained. The above developed conceptual model (*cf.* section. 4.1.2) led to the choice of the different dependent variables. In this respect, three distinct segmentation procedures have been performed on the SME sample dealing with the following dependent variables: (i) the growth of the firm during the considered three-year period; (ii) the introduction of innovation during the same period; and (iii) the existence of interactions with KIBS.

Table 5.1 presents the explicative variables by distinguishing the ones that have been retained and the ones that have not by the three CHAID procedures (performed with Pearson's χ^2 tests at a 5% significance level).

Table 5.1:   **Dependent and explicative variables of the CHAID procedures**

Dependent variables	Explicative variables retained ("best predictors")	Explicative variables not retained
#1 GROWTH	SIZE - SECTOR – LEVEXP	COUNT - REGTYP - IKIBS - PKIBS - NITI - LEVRD -NETSALES - INNOV
#2 INNOV	LEVRD - SIZE - IKIBS – PKIBS – REGTYP	SECTOR - NITI - GROWTH – LEVEXP – COUNT
#3 IKIBS	INNOV - COUNT - NITI – LEVRD	SECTOR - SIZE - REGTYP – GROWTH - NETSALES - LEVEXP

5.1.1   First segmentation

The first segmentation (*cf.* figure 5.1) reveals an important "size effect". The first best predictor indicates that with respect to growth in terms of employment, "size matters", but in a negative way. Dealing with size variation, this result has to be expected and small and medium-sized firms (*i.e.* with less than 99 employees) of the sample show a greater propensity not to reduce and/or to expand their manpower than the bigger ones (with personnel comprised of between 100 and 499). Additionally, some sectoral effects are detected. For the firms comprised of between 20 and 99 employees, sectors such as basic metals and electrical machinery are characterised by positive size variations in comparison with other sectors, particularly textile, wood and machinery manufacturing. The last χ^2 tests-based segmentation concerns the sub-population of medium-sized food industry and chemicals manufacturing firms. For this sub-category (featuring possibly only traditional firms), the less export-oriented firms show a greater propensity for growth than the others. This may seem astonishing, however it concerns only approximately 10 % of the considered population. More generally, and confronting these results with single bivariate tests encompassing the variable "growth" (*cf.* appendix B) an important contribution for the interpretation lies in the variables not selected as best predictors. The fact that variables like "introduction of innovation" or "interaction with KIBS" do not appear selected as best predictors does not mean that there is absolutely no link between them and the propensity of a SME for growth. However, according to these results, if such a link exists, it is strongly subordinated to structural determinants like the size and the sector of activity.

Figure 5.1: "Growth" as a dependent variable

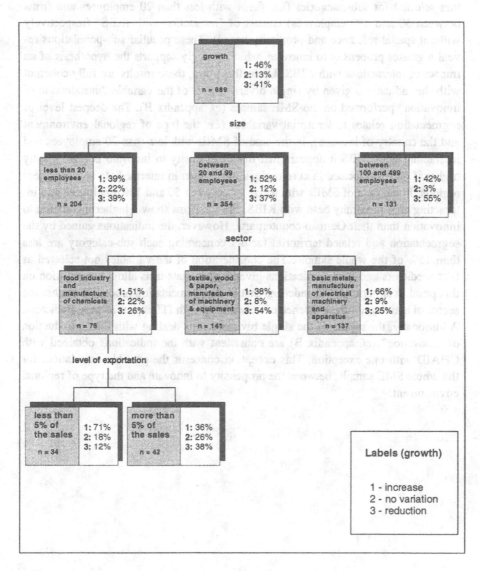

5.1.2 Second segmentation

The second segmentation (*cf.* figure 5.2) stresses the importance of R&D expenses as a best predictor of the variable "introduction of innovation": **firms devoting more than 3,5% of their sales to R&D activities have without exception innovated during the considered period.** The size of firms - in the case of non-R&D intensive SMEs - constitutes the second level of the tree diagram and indicates that the propensity of SMEs to innovate increases with their size. The next best predictors selected for sub-categories (*i.e.* firms with less than 20 employees and firms between 50 and 199 employees) consist of "interaction with KIBS" (respectively without spatial reference and proximity-based). These peculiar sub-populations reveal a greater propensity to innovate, which strongly supports the hypothesis of an impact of interactions with KIBS. Up to this point, these results are fully coherent with the indications given by single bivariate tests of the variable "introduction of innovation" performed on the SME sample (*cf.* appendix B). The deepest level of segmentation relates to territorial variables (*i.e.* the type of regional environment and the country of location). In the case of SMEs with less than 20 employees and interacting with KIBS it appears that their propensity to innovate is significantly higher if they are located in core regions rather than in intermediate and periphery regions. In the case of SMEs with a size of between 50 and 199 employees and interacting on a proximity base with KIBS, French firms show a higher inclination to innovation than their German counterparts. However, the indications gained by the segmentation and related territorial factors concerning each sub-category are less than 15% of the whole sample. The consideration of the variables not selected as best predictors and of the indications given by bivariate tests allow a conclusion on this point. It is important to underline that in the segmentation procedure neither the sector of activity nor the existence of networking with ITI appear as best predictors. Additionally, the results of the single bivariate tests dealing with the "introduction of innovation" (*cf.* appendix B) are consistent with the indications obtained with CHAID with one exception. This exception concerns the possible correlation, for the whole SME sample, between the propensity to innovate and the type of regional environment.

Figure 5.2:   **"Introduction of innovation" as a dependent variable**

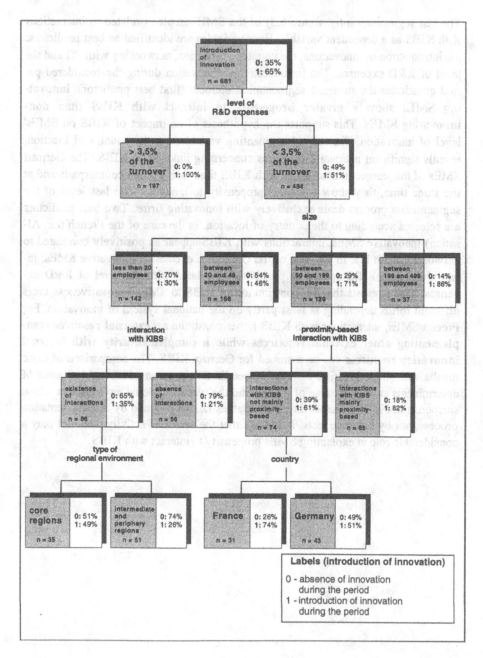

5.1.3   Third segmentation

The last segmentation (*cf.* figure 5.3) of the SME sample is related to interactions with KIBS as a dependent variable. Four variables are identified as best predictors: the introduction of innovations, the country of location, networking with ITI and the level of R&D expenses. The introduction of innovation during the considered period constitutes the strongest segmentation element (first best predictor): **innovating SMEs show a greater propensity to interact with KIBS than non-innovating SMEs.** This supports the hypothesis of an impact of KIBS on SMEs' level of innovation. The second segmenting variable, *i.e.* the country of location, reveals significant national differences concerning linkages to KIBS. The German SMEs of the sample interact more with KIBS than their French counterparts and at the same time, they show a greater propensity to innovate. The last level of the segmentation process deals exclusively with innovating firms. Two best predictors are selected according to the country of location. In the case of the French (*i.e.* Alsatian) innovative SMEs, interactions with KIBS appear as positively correlated to networking with ITI. In the case of the German (*i.e.* Badian) innovative SMEs, interactions with KIBS appear as positively correlated with the level of R&D expenses. This suggests that the contribution of KIBS to SMEs innovativeness takes different forms depending at least partly on the national system of innovation. For French SMEs, interactions with KIBS rather constitute **an external resource complementing other external resources** while **a complementarity with internal innovation resources** can be assumed for German KIBS. The comparison of these results with single bivariate tests (*cf.* appendix B) shows an additional influence of determinants such as the size of a SME and the type of regional environment which surrounds it. Finally, considering the variables neither retained by the segmentation process, nor by bivariate tests, it appears that the sector of activity does not play a considerable role in explaining SMEs propensity to interact with KIBS.

Figure 5.3:   **"Interaction with KIBS" as a dependent variable**

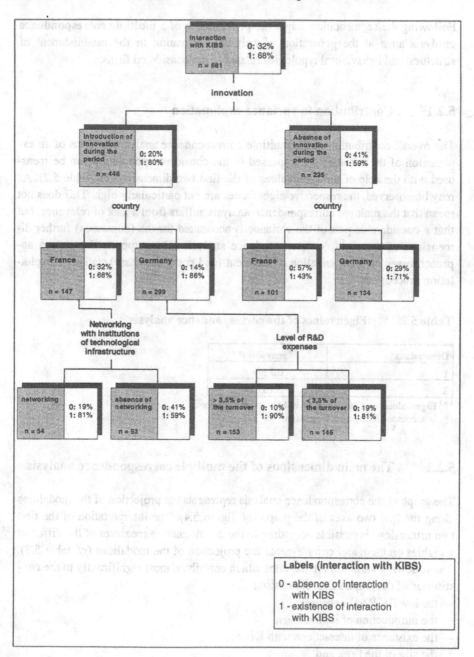

5.2   Multiple correspondence analysis

Following the segmentation stage, the performance of a **multiple correspondence analysis** aims at the generation of helpful information in the establishment of structural and behavioural typologies related to the considered firms.

5.2.1   Contribution to variance explanation

The overall contribution of the multiple correspondence analysis in terms of an explanation of the variance encompassed in the considered population can be measured with the help of the eigenvalues of the first two dimensions (*cf.* table 5.2). As may be observed, the respective eigenvalues are not particularly high. This does not mean that the multiple correspondence analysis suffers from a lack of relevance, but that a considerable part of the variance is accounted for the (numerous) further dimensions of the multiple correspondence analysis. This supports the chosen approach based on a combination of different (and complementary) statistical exploitation procedures.

Table 5.2:   **Eigenvalues of the correspondence analysis**

Dimensions	Eigenvalues*
1	0,2299
2	0,1633

(*) Eigenvalues are a measurement of how much variance
   is accounted for by each dimension.

5.2.2   The main dimensions of the multiple correspondence analysis

The graph of the correspondence analysis represents the projection of the modalities along the first two axes of the graph (*cf.* figure 5.4). The interpretation of the first two dimensions is possible according to the discrimination measures of the different variables on them and complements the projection of the modalities (*cf.* table 5.3). The following variables are the ones which contribute most significantly to the constitution of the graph's first dimension:
- the level of R&D expenses,
- the introduction of innovations,
- the existence of interactions with KIBS,
- the size of the firm, and
- the existence of networking with ITI.

Basically, the first dimension may be seen as opposing the firms which are R&D-intensive, innovative, interacting with KIBS, relatively bigger and networking with ITI (eastern part of the graph) to their less R&D-intensive, non innovative, non interactive, relatively smaller and non networking counterparts (western part of the graph). In the same way, the country of location, the type of regional environment and the sector of activity constitute the variables which contribute the most to the constitution of the second dimension of the graph. This dimension contrasts firms principally located in Germany and in periphery regions, mainly from the basic metal, wood and associated industries (northern part of the graph) to firms situated in France and in core and intermediate regions and whose production relates to the machinery, chemicals textiles or food sectors (southern part of the graph).

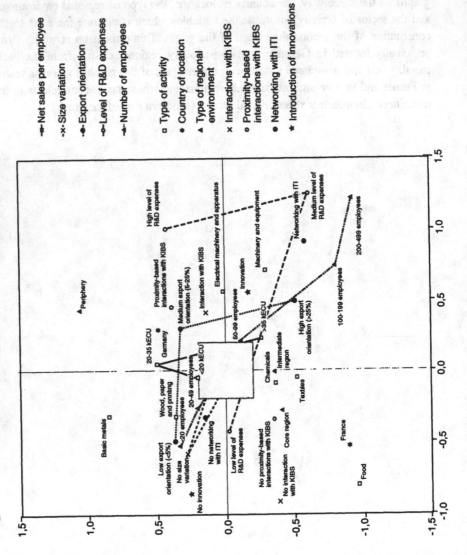

Figure 5.4: Multiple correspondence analysis of the SMEs sample

- ■ Net sales per employee
- × Size variation
- ● Export orientation
- ◇ Level of R&D expenses
- ✦ Number of employees
- □ Type of activity
- ✱ Country of location
- ▲ Type of regional environment
- × Interactions with KIBS
- ○ Proximity-based interactions with KIBS
- ● Networking with ITI
- ✶ Introduction of innovations

Table 5.3:   Discrimination measures of the correspondence analysis

Variables	Discrimination measures* on dimension 1	Discrimination measures* on dimension 2
COUNT	0,156	0,441
LEVRD	0,470	0,075
SECTOR	0,238	0,285
SIZE	0,308	0,200
REGTYP	0,075	0,393
INNOV	0,423	0,040
IKIBS	0,387	0,061
NITI	0,297	0,108
LEVEXP	0,176	0,130
PKIBS	0,164	0,131
NETSALES	0,012	0,077
GROWTH	0,054	0,020

(*) Discrimination measures indicate the variables' contribution to the constitution of the axes.

5.2.3   Interpretation of the correspondence analysis

The following information can been gain from the multiple correspondence analysis. Firstly, the constitution of the correspondence analysis is mainly influenced by: (i) the intensity of R&D activities; (ii) the introduction of innovations; and (iii) by the existence of interactions with KIBS. In fact, the analysis quite strongly contrasts two types of behaviours, additionally revealing the role played by interactions with KIBS and by networking with ITI. This supports the idea that firms having introduced innovation performed specific efforts in order to do so, that internal and external innovation efforts are linked and that interactions with KIBS are linked to these efforts.

Secondly, there is no clear indication related to the economic results of the considered firms. Considering the graph of the modalities it is for instance not possible to associate directly "growing firms" with "innovating an interacting firms". On the other hand, this does not mean that the former are necessarily less successful than non "innovating and/or interacting" ones, but only that the reality is more complex and that other dimensions should be taken into account. This is strongly suggested by the variables GROWTH and NETSALES which "turn around" more dimensions than the axes 1 and 2. Additionally, considering "structural" determinants, the number of employees (or size of the firm) appears to be particularly determining. However, the absence of a strong (and linear) correlation between LEVRD and SIZE suggests the existence of a "critical mass" effect: R&D expenses seem to increase with the size of the firms, but only up to a certain level. In other words, the most

R&D intensive SMEs are not the smaller ones, but also not the biggest ones. Considering location variables, a clear opposition can be observed related to the country of location. This reflects divergent behaviours, notably concerning interactions with KIBS: German SMEs interact more with KIBS than their French counterparts. The opposition is less sharp when examining the type of regional environment. In fact, a certain proximity can be observed between the modalities "core" and "intermediate", in opposition to "periphery".

Finally, the multiple correspondence analysis shows an important sectoral divergence. In this respect, and according to the first dimension ("innovativeness and/or interactivity") two sub-groups may be distinguished: (i) firms which are stronger R&D/innovation/interaction oriented associating machinery, equipment, electrical machinery and apparatus; in contrast with (ii) firms that are less R&D/innovation/ interaction oriented regrouping the other sectors (*i.e.* mainly the manufacturing of food products, of textiles, of wood and paper, of chemicals and of basic metals). In conclusion, the multiple correspondence analysis clearly completes the elements delivered by the segmentation procedure. Moreover, since the multiple correspondence analysis gives indication in terms of typology (or modalities clustering) it generates valuable information for the path modelling step.

5.3   Path modelling

The path modelling procedure requires the reduction of the considered variables to dichotomic ones. The results relate to the significance values of the relations tested and are presented in the form of a graph depicting the "paths" identified as significant.

5.3.1   Variables reduction

Path modelling is based on a "cascade combination" of PROBIT algorithms. Since PROBIT procedures can only be performed for dichotomic dependant variables, it was necessary to "binarise" the variables of the set (*cf.* table 5.4). This reduction of modalities to a 0/1 scheme constitutes a loss of information. However, the combination and merging of categories derives from the results of the segmentation and correspondence analyses so there is no oversimplification. On the contrary, PROBIT algorithms detect dependencies that could not be observed in the first (CHAID) and second (multiple correspondence analysis) steps of the empirical investigation.

Table 5.4:   Set of dichotomic variables used for the path-modelling

Variable code	Variable description	Labels	
GROWTH	*Variation in the number of employees (1992-1995)*	0- 1-	Reduction or no variation Increase
NETSALES	*Net sales per employee (1995)*	0- 1-	< 35 KECU per capita > 35 KECU per capita
LEVEXP	*Export orientation (1995)*	0- 1-	Weakly export oriented (< 25 % of the turnover) Strongly export oriented (> 25 % of the turnover)
INNOV	*Performance of innovations (1992-1995)*	0- 1-	Absence of innovation during the period Introduction of innovation during the period
IKIBS	*Interaction with KIBS*	0- 1-	Absence of interactions with KIBS Existence of interactions with KIBS
PKIBS	*Proximity with KIBS*	0- 1-	Interactions with KIBS not mainly proximity-based Interactions with KIBS mainly proximity-based
NITI	*Networking with ITI*	0- 1-	Absence of networking with ITI Existence of networking with ITI
LEVRD	*Level of R&D expenses*	0- 1-	Low (R&D expenses < 3,5% of the turnover) Medium or high (R&D expenses > 3,5% of the turnover)
SECTOR	*Type of activity*	0- 1-	Manufacture of food products; of textiles; of wood, paper and printing; of chemicals; of basic metals Manufacture of machinery and equipment; of electrical machinery and apparatus
SIZE	*Number of employees (1995)*	0- 1-	10 to 99 employees 100 to 499 employees
COUNT	*Country of location*	0- 1-	Germany France
REGTYP	*Type of regional environment*	0- 1-	Periphery region Core and intermediate regions

5.3.2   Results

The principle of the path modelling consists of successive tests of the explanatory character of different variables in relation to different dependent variables. This was made according to the conceptual model developed in chapter 4. The results of the different PROBIT algorithms are shown in table 5.5 (*cf.* also appendix C) and the "path diagram" constitutes its graphic translation (*cf.* figure 5.5).

Table 5.5: **Results of the PROBIT analysis**

Dependent variables [prob > χ^2 of the model tested]	Variables presenting significance values < 0,01 according to PROBIT estimates	Variables presenting significance values between 0,01 and 0,05 according to PROBIT estimates	Variables presenting significance values > 0,05 according to PROBIT estimates
GROWTH [0,1625]	-	-	LEVEXP – NETSALES - INNOV - LEVRD - IKIBS - PKIBS - NITI – SIZE -COUNT – REGTYP – SECTOR
NETSALES [0,0008]	LEVEXP - IKIBS	-	GROWTH – INNOV - LEVRD - PKIBS - NITI - SIZE - COUNT – REGTYP – SECTOR
LEVEXP [0,0000]	NETSALES - SECTOR	-	GROWTH – INNOV - LEVRD - IKIBS - PKIBS - NITI - SIZE – COUNT - REGTYP
INNOV [0,0000]	LEVRD* - IKIBS - NITI – SIZE	SECTOR	PKIBS - COUNT - REGTYP
LEVRD [0,0000]	IKIBS - NITI - SECTOR	SIZE – COUNT	PKIBS – REGTYP
IKIBS [0,0000]	NITI - COUNT	SIZE	REGTYP – SECTOR
PKIBS [0,0061]	COUNT	-	NITI - SIZE – REGTYP – SECTOR
NITI [0,0000]	IKIBS - PKIBS - SIZE – SECTOR	-	COUNT – REGTYP

* *LEVRD: variable dropped out of the analysis after the first iteration of the PROBIT algorithm since it predicts INNOV perfectly.*

Figure 5.5:   **Path modelling: the SMEs sample**

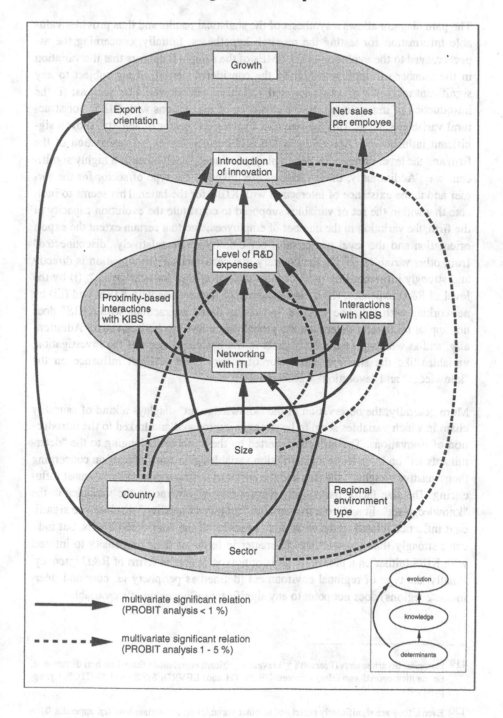

5.3.3   Interpretation

The path diagram allows a synthesis of the statistical results and thus provides valuable information for testing the research hypotheses. Initially, concerning the aspects related to the evolution of the SMEs of the sample it appears that the variation
in the number of employees (during the considered period) is not subject to any
significant relation with other selected variables. Neither variables such as: (i) the
introduction of innovations; (ii) the existence of interactions with KIBS; nor structural variables such as: (iii) the size of a firm; or (iv) sector of activity, show a significant influence.[119] According to PROBIT results, the export orientation of the
firm and the level of net sales per employee influence each other in a highly significant way. Additionally, these variables are linked to the type of sector for the former and to the existence of interaction with KIBS for the latter. This seems to indicate that within the set of variables supposed to constitute the evolution capacity of
the firm, the variation in the number of employees, and to a certain extent the export
orientation and the level of net sales per employee, are relatively "disconnected"
from other variables.[120] On the contrary, the performance of innovation is directly
and strongly influenced by most of the variables of the "knowledge set": (i) by the
level of R&D expenses; (ii) by the existence of interactions with KIBS; and (iii) by
networking with ITI. The variable "proximity-based interactions with KIBS" does
not appear to directly determine the probability a SME to be innovative. Additionally, and as was expected with regards to the precedent steps of the investigation,
variables like the size and the sector of a firm exert a direct influence on the
"knowledge" and "evolution" variables sets.

More generally, the observation of the "knowledge set" displays a kind of causality
chain in which variables seem to be strongly interrelated and linked to the introduction of innovations. The influence exerted by the variables belonging to the "determinants set" on knowledge and evolution variables give some indication concerning
their "relative weights": the size and the sector of a firm appear to be the most influencing. The size factor particularly plays a role in influencing the variables in the
"knowledge set". In addition, the variable "country of location" discloses no significant influence related to the evolution capacities of the considered SMEs, but indicates strongly that national systems matter in terms of their propensity to interact
with KIBS (either on a proximity basis or not) as well as in terms of R&D intensity.
Finally, the type of regional environment (defined as periphery *vs.* core and intermediate regions) does not point to any significant influence on other variables.

---

119 However, bivariate tests (Pearson's χ^2) reveal significant correlations (based on non-dichotomic,
    i.e. multicategorial variables) between GROWTH and: LEVRD, SIZE and SECTOR (*cf.* ap-
    pendix B).

120 Even if they are significantly correlated to other variables on a bivariate base (*cf.* appendix B).

To summarise the results related to SMEs, it is possible to argue that for the firms of the examined samples, the overall notion of "evolution" covers quite diversified situations. More precisely, for the considered three-year period, no direct link can be drawn between the introduction of innovations and the variation in the size of those firms or their economic performance. This does not mean that such influences do not exist, they may be detected on a bivariate base (*cf.* appendix B), however they are not highly significant. More convincing in statistical terms is the influence of KIBS on SMEs' innovativeness (and economic performance). This direct impact is reinforced indirectly by the simultaneous activation of additional internal and external innovation sources. This suggests that interactions with KIBS constitute for SMEs a complementary rather than a substitutable innovation resource. Finally, it is important to stress the variables exerting, in the frame of the multivariate analysis, limited or no influence on other variables. In this respect, the type of regional environment surrounding a SME seems neither to induce particular behaviour nor to generate specific consequences in terms of firm evolution. This suggest that, at least for the samples and territorial units considered, "region does not matter". Additionally, proximity-based interactions with KIBS only significantly influence the propensity a SME to network with ITI.

Conclusion

As a result of the empirical examination, the following elements can be assumed concerning the evolution patterns of the observed SMEs. Firstly, there is statistical evidence showing the impact of interactions with KIBS on the innovation capacities and behaviour of SMEs. Secondly, even if the economic performance of SMEs is not directly affected by their innovation effort, they are determined at least partially by some elements constituting the knowledge base of the considered firms, notably by the existence of interactions with KIBS. Finally, the statistical investigation allows the perception of the respective influences of structural and territorial factors determining innovation and interaction patterns of SMEs. The parallel investigation of the KIBS sample, presented in the next chapter, adopts the same methodology. This will allow the comparison of the findings related to SMEs and KIBS with the results of other innovation surveys in the final chapter.

Chapter 6: Statistical exploitation of the KIBS sample

Introduction

In this chapter, the KIBS sample is investigated using the same methods as for the SME sample, *i.e.* the statistical analysis combines segmentation procedures, multiple correspondence analysis and PROBIT algorithms.

6.1 Segmentation procedures

As was the case for the SME sample, the first empirical step is the application of **segmentation algorithms (CHAID)**. This procedure, as shown in section 4.3.1, successively detects correlations between variables. Three distinct segmentation procedures have been performed according to the conceptual model. The dependent variables considered are: (i) the growth of the firm during the considered three-year period; (ii) the introduction of innovations during the same period; and (iii) the existence of interactions with SMEs.

The explicative variables which have been respectively retained and not retained by the three CHAID procedures (performed with Pearson's χ^2 tests at a 5% significance level) are presented in table 6.1.

Table 6.1: Dependent and explicative variables of the CHAID procedures

Dependent variables	Explicative variables retained ("best predictors")	Explicative variables not retained
#1 GROWTH	SIZE - ISMES - INNOV - COUNT	SECTOR – REGTYP - PSMES - NITI - LEVRD – TURNOVER – LEVEXP
#2 INNOV	LEVRD - SIZE - GROWTH - PSMES	SECTOR – COUNT – REGTYP - ISMES - NITI – TURNOVER – LEVEXP
#3 ISMES	LEVRD - SECTOR - GROWTH - LEVEXP	SIZE – COUNT – REGTYP - NITI - TURNOVER – INNOV

6.1.1 First segmentation

Considering the variation in the number of employees during the 1992-1995 period, the first segmentation (*cf.* figure 6.1) shows the importance of the variable "size of the firm" in selecting it as the first "best predictor". As has been underlined in the case of the SME sample, "size matters" with respect to growth. However, the phenomenon here is the opposite of what has been observed by SMEs: larger KIBS (especially with 5 or more employees) show a higher propensity to increase their size than smaller ones.[121] The next "best predictors" selected by the segmentation algorithm consist of the "existence of interactions with SMEs" in the case of KIBS smaller than 3 employees and of the "introduction of innovations" in the case of KIBS with 3 or more employees respectively. This clearly indicates that for the respective considered sub-population of firms, growing KIBS may be characterised as being more interactive with SMEs and more innovating than the non-growing ones. Finally, a further segmentation step is provided for the sub-population of non-interacting KIBS with less than 3 employees. In fact, it appears that the country of location significantly influences the variation in size of the considered KIBS and indicates a more important stability in the case of French firms. The confrontation of these results with single bivariate tests encompassing the variable "growth" (*cf.* appendix B) show, in addition, a statistically significant influence (at 5% level) of R&D investment on KIBS' propensity for growth, which is not retained by the CHAID procedure. Finally , it is interesting to note that the type of activity (variable SECTOR) has not been selected as best predictor, which indicates that KIBS' growth behaviour does not vary significantly across sectors.

---

121 Nevertheless, it should be borne in mind that the size categories for SMEs and KIBS are far from being identical. For instance, "small" KIBS have less than 3 employees whereas "small" SMEs have less than 20 employees.

Figure 6.1: "Growth" as a dependent variable

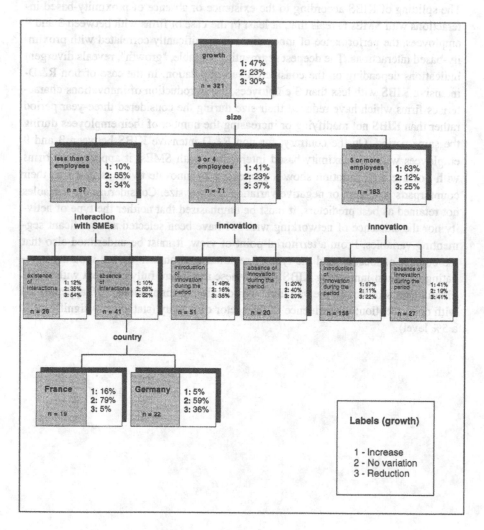

6.1.2 Second segmentation

The second segmentation (*cf.* figure 6.2) demonstrates the importance of internal R&D as best predictor for the introduction of innovations in an identical way as to what has been observed for SMEs: firms devoting more than 3,5% of their turnover to R&D activities have all introduced innovations during the considered period. For the other firms (the non R&D-intensive KIBS) the next best predictor is the size. The segmentation clearly indicates that the propensity to introduce innovations increases dramatically with the number of employees. Two additional variables are detected by the algorithm as further predictors: proximity-based interactions with

SMEs and the variation in the number of employees during the considered period. The splitting of KIBS according to the existence or absence of proximity-based interactions with SMEs reveals that, at least in the case of firms with between 3 and 9 employees, the performance of innovations is significantly correlated with proximity-based interactions. The deepest segmenting variable, "growth", reveals divergent indications depending on the considered sub-population. In the case of non R&D-intensive KIBS with less than 3 employees, the introduction of innovations characterises firms which have reduced their size during the considered three-year period rather than KIBS not modifying or increasing the number of their employees during the same period. On the contrary, for non R&D-intensive KIBS between 3 and 9 employees without proximity-based interactions with SMEs, it appears that firms with positive size variation show a propensity to innovate twice as much as their counterparts with zero or negative variation in their size. Considering the variables not retained as best predictors, it must be emphasised that neither the type of activity nor the existence of networking with ITI have been selected as significant segmenting variables. From a territorial point of view, it must be underlined also that the country of location and the type of regional environment do not separate innovating from non-innovating KIBS. All of these results are fully coherent with single bivariate tests related to the variable "introduction of innovations" (*cf.* appendix B) with one exception: the influence of the sector of activity (statistically significant at a 5% level).

Figure 6.2: **"Performance of innovation" as a dependent variable**

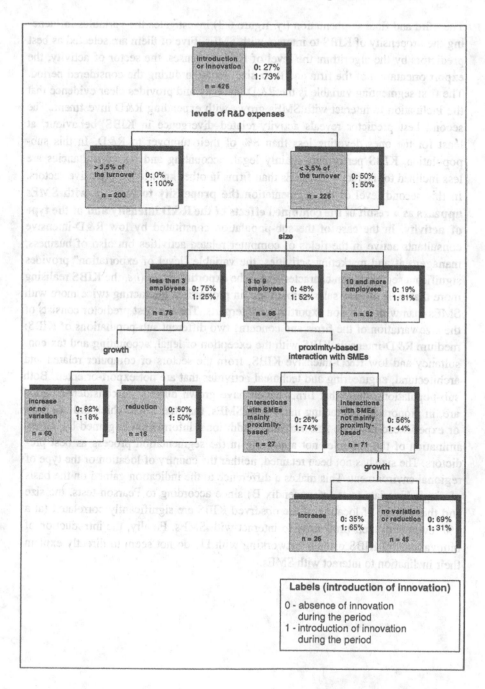

6.1.3   Third segmentation

The third and final segmentation (*cf.* figure 6.3) investigates the variables influencing the propensity of KIBS to interact with SMEs. Five of them are selected as best
predictors by the algorithm: the level of R&D expenses, the sector of activity, the
export orientation of the firm and their size variation during the considered period.
The first segmenting variable is the R&D intensity and provides clear evidence that
the inclination to interact with SMEs grows with expending R&D investment. The
second best predictor reveals activity-related divergence in KIBS' behaviour, at
léast for the ones devoting less than 8% of their turnover to R&D. In this sub-
population, KIBS performing mainly legal, accounting and tax consultancies are
less inclined to interact with SMEs than firms in other knowledge-intensive sectors.
In this second level of the segmentation **the propensity to interact with SMEs
appears as a result of the combined effects of the R&D intensity and of the type
of activity.** In the case of the sub-population constituted by low R&D-intensive
consultants active in the fields of computer related activities but also of business,
management and marketing activities, the variable "level of exportation" provides
significant segmentation characteristics. The exporting firms (*i.e.* the KIBS realising
more than 5 % of their sales abroad) are, in proportion, interacting twice more with
SMEs than with their non exporting counterparts. The last best predictor consists of
the size variation of the firms and concerns two different sub-populations of KIBS:
medium R&D-intensive KIBS with the exception of legal, accounting and tax consultancy and low R&D-intensive KIBS, from the sectors of computer related and
architectural, engineering and technical activities that are not export-oriented. Both
sub-populations show that firms which have grown during the considered period
are, in proportion, interacting more with SMEs, than the firms which did not grow
or experienced negative size variation. Additional information is gained by the examination of the variables not appearing in the segmentation process as best predictors. The size has not been retained, neither the country of location or the type of
regional environment. This makes a difference to the indication gained on the basis
of single bivariate tests (*cf.* appendix B) since according to Pearson-tests, the size
and the country of location of the observed KIBS are significantly correlated (at a
1% level) with their propensity to interact with SMEs. Finally, the introduction of
innovations by KIBS or their networking with ITI do not seem to directly explain
their inclination to interact with SMEs.

Figure 6.3:   **"Interaction with SMEs" as a dependent variable**

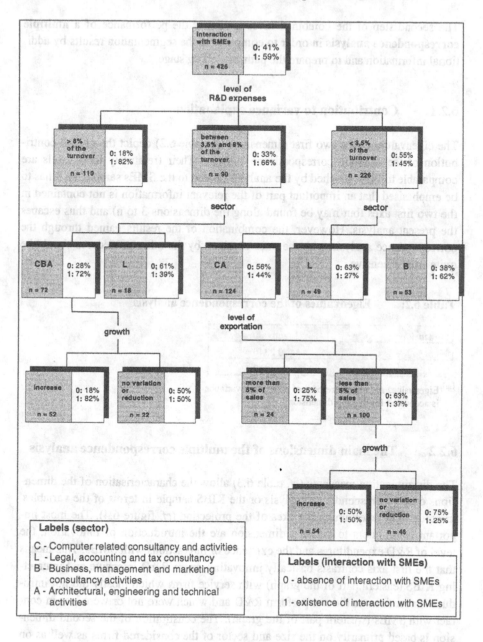

6.2   Multiple correspondence analysis

The second step of the combined investigation is the performance of a **multiple correspondence analysis** in order to complement the segmentation results by additional information and to prepare the path-modelling stage.

6.2.1   Contribution to variance explanation

The eigenvalues of the two first dimensions (*cf.* table 6.2) depict the overall contribution of the multiple correspondence analysis. Their (relatively low) levels are comparable to those reached by the analysis related to the SMEs sample, so it has to be emphasised that an important part of the relevant information is not contained in the two first axes (but may be found along the dimensions 3 to n) and thus escapes the present analysis. However, the combination of the results gained through the correspondence analysis with the ones obtained by the other procedures should be quite comprehensive.

Table 6.2:   **Eigenvalues of the correspondence analysis**

Dimensions	Eigenvalues*
1	0,2340
2	0,1417

(*) Eigenvalues are a measurement of how much variance
   is accounted for by each dimension.

6.2.2   The main dimensions of the multiple correspondence analysis

The discrimination measures (*cf.* table 6.3) allow the characterisation of the dimensions of the correspondence analysis or the KIBS sample in terms of the variables' importance in determining the axes of the projection (*cf.* figure 6.4). The most important contributors to the first dimension are the introduction of innovation, the level of R&D expenditures and the existence of interactions with SMEs. This means that the first axis contrasts basically innovating, more R&D intensive and interacting KIBS (eastern part of the graph) with service firms which mostly did not introduce innovations or did not perform R&D and which were not or were less in contact with SMEs (western part of the graph). The constitution of the second dimension is based primarily on the size and sector of the considered firms as well as on the existence of proximity-based interactions with SMEs. This has as a consequence that the graph may be seen as facing its northern part (characterised by bigger firms, no or only limited interactions with SMEs on a proximity base and typically being active as legal, accounting or tax consultants) to its southern part (constituted by smaller KIBS, typically doing business, management or marketing consultancy and characterised by a higher propensity to interact with SMEs on a proximity base).

Table 6.3:   **Discrimination measures of the correspondence analysis**

Variables	Discrimination measures* on dimension 1	Discrimination measures* on dimension 2
SIZE	0,304	0,378
LEVRD	0,501	0,155
ISMES	0,459	0,150
INNOV	0,558	0,037
PSMES	0,208	0,215
SECTOR	0,141	0,246
GROWTH	0,219	0,157
LEVEXP	0,130	0,127
REGTYP	0,017	0,167
NITI	0,130	0,018
COUNT	0,113	0,008
TURNOVER	0,029	0,043

(*) Discrimination measures indicate the variables' contribution to the constitution of the axes.

Figure 6.4:   Multiple correspondence analysis of the KIBS sample

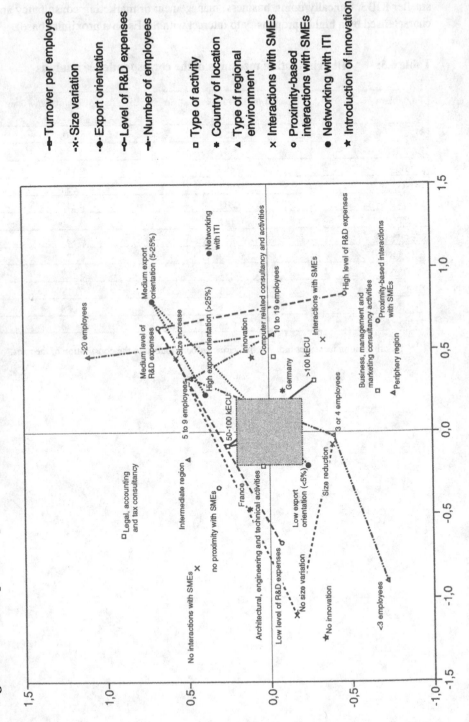

6.2.3   Interpretation of the correspondence analysis

The multiple correspondence analysis of the KIBS sample is particularly helpful for gaining complementary knowledge on three levels. Firstly, the variables with the highest contribution to the constitution of the first axis clearly indicate the importance of innovation and R&D for a typology of KIBS. A substantial complementary element consists of the additional role of interactions with SMEs. This means that the firms which are supporting SME innovations are also, typically, investing (through R&D) in their own development and introducing (internal) innovation.

The second aspect to emphasise concerns the elements gained in terms of size and sector typology. In fact, it is possible to associate the innovative/interactive behaviour[122] with KIBS coming typically from the business, management and marketing consultancy sector and to a certain extent from computer related activities. Those innovative/interactive KIBS mainly employ 3 to 20 persons so that they do not correspond either to the smallest or to the biggest size categories. On the contrary, non innovative/interactive KIBS may be seen as typically either very small (less than 3 employees) or very large (more than 20 employees) for KIBS characteristics. In terms of sector characterisation, according to the correspondence analysis they are more frequently active in the fields of legal, accounting and tax consultancy. The firms from the architectural, engineering and technical activity sectors may also be associated with this feature, but with weaker intensity.

The final important elements to note are related to economic performance and to the importance of territorial determinants. Concerning the possible relation between variables featuring economic performance (GROWTH, LEVEXP, TURNOVER) and the introduction of innovations or the existence of interactions with SMEs, no clear indication can be found even if – at least intuitively – the projection suggests some links (for instance related to the level of turnover per employee). Likewise, in the frame of the correspondence analysis, no particular strong contribution can be attributed to territorial determinants (with the exception of proximity effects in KIBS/SMEs interactions)

In conclusion, it is important to underline that the information gained in the second step of the analysis, will be used under the form of redefinition of variables for the path modelling which is the next stage of the empirical investigation for the KIBS sample.

---

[122] Corresponding to the south-eastern part of the projection, this innovative/interactive feature comprises firms with one or more of the following characteristics: (i) the introduction of innovation during the period concerned; and/or (ii) a high level of R&D expenditure; and/or (iii) innovation-related interactions with SMEs; and/or (iv) proximity-based interactions with SMEs.

6.3   Path modelling

The same procedure as the one employed for the SME sample (*cf.* section 5.3) is followed here: the variables are binarised in order to allow the performance of PROBIT algorithms. The results of the different analyses are presented in the form of a graph depicting the "paths" identified as significant.

6.3.1   Variable reduction

As required for the SMEs sample, it is also necessary here to "reduce" the variables to dichotomic ones in order to allow PROBIT algorithms to be carried out. These dichotomic variables are presented in table 6.4.[123] They have been constituted on the basis of the clustering of modalities detected in the previous analytical step, *i.e.* the multiple correspondence analysis.

Table 6.4:   Set of dichotomic variables used for the path-modelling

Variable code	Variable description	Labels	
GROWTH	*Variation in the number of employees (1992-1995)*	0-	Reduction or no variation
		1-	Increase
TURNOVER	*Turnover per employee (1995)*	0-	< 100 KECU
		1-	> 100 KECU
LEVEXP	*Export orientation (1995)*	0-	Not export oriented (less than 5 % of the turnover)
		1-	Export oriented (more than 5 % of the turnover)
INNOV	*Performance of innovations (1992-1995)*	0-	Absence of innovation during the period
		1-	Introduction of innovation during the period
ISMES	*Interactions with SMEs*	0-	Absence of interactions with SMEs
		1-	Existence of interactions with SMEs
PSMES	*Proximity to SMEs*	0-	Interactions with SMEs not mainly proximity-based
		1-	Interactions with SMEs mainly proximity-based
NITI	*Networking with ITI*	0-	Absence of networking with ITI
		1-	Existence of networking with ITI
LEVRD	*Level of R&D expenses*	0-	R&D expenses < 3,5 % of the turnover
		1-	R&D expenses > 3,5 % of the turnover
SECTOR	*Type of activity*	0-	Legal, accounting and tax consultancy; architectural, engineering and technical activities
		1-	Computer related consultancy and activities; business, management and marketing consultancy activities
SIZE	*Number of employees (1995)*	0-	Less than 5 employees
		1-	More than 5 employees
COUNT	*Country of location (region)*	0-	Germany
		1-	France
REGTYP	*Type of regional environment*	0-	Periphery region
		1-	Core or intermediate region

---

123 The detailed results are shown in appendix C.

6.3.2   Results

According to the conceptual model developed in chapter 4, and thus respecting the investigation structure used for the SMEs sample, several PROBIT algorithms have been carried out. The results are presented in table 6.5. The "path diagram" (*cf.* figure 6.5) illustrates the combination of the different results.

Table 6.5:   Results of the PROBIT analysis

Dependent variables [prob > χ^2 of the model tested]	Variables presenting significance values < 0,01 according to PROBIT estimates	Variables presenting significance values between 0,01 and 0,05 according to PROBIT estimates	Variables presenting significance values > 0,05 according to PROBIT estimates
GROWTH [0,0000]	INNOV - SIZE	-	LEVEXP - TURNOVER - LEVRD - ISMES - PSMES - NITI - COUNT - REGTYP - SECTOR
TURNOVER [0,0122]	SECTOR	-	LEVEXP - GROWTH - INNOV - LEVRD - ISMES - PSMES - NITI - SIZE - COUNT - REGTYP
LEVEXP [0,0000]	ISMES - PSMES	SIZE - SECTOR	GROWTH - TURNOVER - INNOV - LEVRD - NITI - COUNT - REGTYP
INNOV [0,0000]	LEVRD[1] - NITI[1] - SIZE	PSMES	ISMES - COUNT - REGTYP - SECTOR
LEVRD [0,0000]	ISMES - COUNT	NITI	PSMES - SIZE - REGTYP - SECTOR
ISMES [0,0000]	SIZE - SECTOR	-	NITI - COUNT - REGTYP
PSMES [0,1033][2]	_[2]	_[2]	[NITI - SIZE - COUNT - REGTYP - SECTOR] [2]
NITI [0,0009]	COUNT	-	ISMES - PSMES - SIZE - REGTYP - SECTOR

[1] *LEVRD and NITI: variables dropped out of the analysis after the first iteration of the PROBIT algorithm since they predict INNOV perfectly.*

[2] *prob > χ^2 = 0,1033 indicates that the model tested with PSMES as a dependent variable and NITI, SIZE, COUNT, REGTYP and SECTOR as independent variables is not significant.*

Figure 6.5:   Path modelling: the KIBS sample

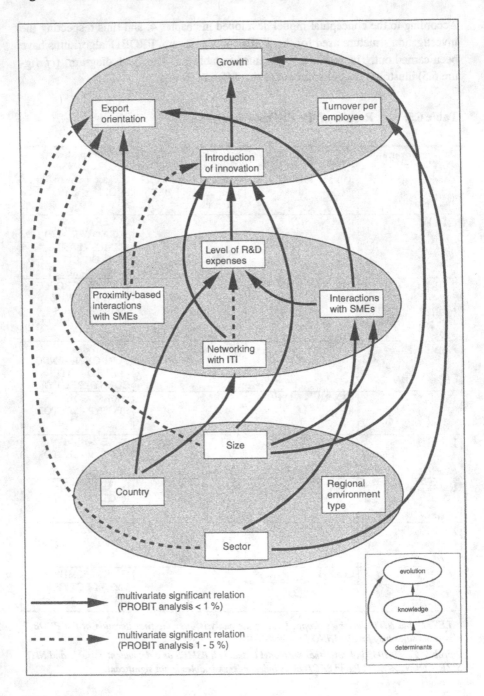

6.3.3 Interpretation

Considering the results of the PROBIT analyses and the ensuing path modelling, two main findings can be advanced concerning the structural relations within the conceptual model. Firstly, the variables constituting the "evolution set" of the model are influenced by knowledge-base related variables as well as by variables corresponding to structural determinants (*i.e.* the size of the firm and its sector of activity). It has to be stressed that variations in size appear as significantly linked to the introduction of innovations: innovating KIBS show a greater propensity to grow than non-innovating ones. Concerning the level of turnover per employee, this sector of activity is the sole variable revealing a strong impact: for instance, management-oriented or computer related consultancies show levels of turnover per employee that are significantly higher than other KIBS' activities. The level of exportation, on the contrary, seems to depend significantly on several variables, notably on the existence of interactions with SMEs, both without spatial reference as well as on a proximity base, and – with a lower significance – on size and sector of activity. The variable "introduction of innovations" shows an even more complex structure in terms of influences. In fact, it is affected by: (i) internal resources (the level of R&D expenditures); (ii) by external factors (networking with ITI and proximity-based interactions with SMEs); and (iii) by the size of the firm (the propensity to innovate increases with the size[124]).

The second point is related to the variables featuring the knowledge base of the firm. The level of R&D expenditures is significantly correlated to the country of location (German KIBS invest significantly more in R&D than their French counterparts), but depends also on interactions with SMEs as well as on networking with ITI. Those interactions themselves appear to be strongly linked to the firms' structural characteristics (size and sector of activity).[125] On the contrary, the analysis does not reveal significant impacts of structural determinants explaining proximity-based interactions with SMEs.[126] Finally, concerning the "knowledge" set of variables, the inclination of KIBS to co-operate with ITI is strongly influenced by the country of location (KIBS in Alsace interact significantly less with ITI than their counterparts in Baden).

The path modelling procedure identified elements that are helpful for explaining the behaviour of the considered KIBS. Firstly, there is clear empirical evidence in fa-

---

[124] This size effect can be observed at least until a firm's critical size of about 20 employees as it is has been established with the help of the multiple correspondence analysis (*cf.* section 6.2.3).

[125] As well as the country of location, but only in bivariate tests (*cf.* appendix B).

[126] KIBS' size and sector appear as factors influencing the propensity to interact, on the basis of proximity, with SMEs but only when considering bivariate tests (*cf.* appendix B) and not when examining multivariate relations.

vour of a direct (statistical) relation between innovative behaviour and positive size variation. Additionally, indirect effects should be taken into account. For instance, the existence of interactions (whether proximity-based or not) with SMEs appears to have a positive influence on KIBS' propensity to export their services. In the same way, variables like the level of R&D investments or the existence of networking activities with ITI are liable to have an impact on the growth rate via their impact on innovation activities. However, it is necessary to underline that especially the level of turnover per employee, which is typically an indicator of economic performance rather than of firm evolution, does not seem to be strongly influenced by innovativeness or even by any knowledge-related variable. Secondly, the expected direct link between interactions with SMEs and innovative behaviour cannot be supported with strong empirical evidence.[127] Nevertheless, there are clear indications that KIBS' innovations and interactions with SMEs are not independent. At least, there are indirect effects in favour of a positive influence of interactions with SMEs on the innovativeness of KIBS notably via their R&D expenses. Finally, considering structural and locational variables, it appears that the size of the firm and the sector of activity rather have an impact on variables from the "evolution set" than on the ones related to the knowledge base aspects. On the contrary, the country of location only influences the last category of variables (*i.e.* the R&D intensity and the contacts with KIBS) whereas the type of regional environment does not significantly affect any variable.

Conclusion

This chapter, focussing on the empirical investigation of the KIBS sample, clearly shows that the evolution capacity of a KIBS is strongly determined by the contents and structure of its knowledge base. Additionally, and in particular if one considers innovation processes, it appears that KIBS' evolution patterns are also affected by structural factors (such as the size and, to a certain extent, the type of activity) and are subject to territorial influences. The findings support the hypothesis of the positive impact of interactions with SMEs on KIBS' innovativeness and more generally, the main empirical results can be summarised with the help of the following three points. Firstly, the investigation provides statistical evidence concerning the positive impact of interactions with SMEs on KIBS' innovation capacities. Secondly, the statistical analysis establishes the influence on a KIBS' economic performance of: (i) the introduction of innovations within a KIBS; as well as (ii) the existence of interactions with SMEs. Finally, concerning the impact of territorial factors, the empirical examination of the KIBS' sample highlights the importance of proximity-

---

127 With the exception of the influence of proximity-based interactions with SMEs which are significant at a 5% level in a multivariate way.

based interactions with SMEs, the inescapable influence of national innovation systems but stresses the lack of evidence in regard to the impact of the type of regional environment. In the next chapter, the key findings related to the KIBS sample are arranged parallel to the analysis completed for the SME sample and are confronted with empirical results extracted from other surveys in order to allow the formulation of theoretical and policy-related implications.

Chapter 7:   Main findings and policy implications

Introduction

The final chapter attempts to present a general interpretation of the key findings related to SMEs and KIBS, to provide a broader scope to the analysis and to draw policy relevant conclusions from the whole investigation. In this respect, the results of the investigation are synthesised and discussed in the light of other studies in section 7.1. Then, adopting a broader perspective, an integrated typology of innovation interactions involving SMEs and KIBS is proposed in section 7.2. Finally, the main policy implications of the research are discussed in section 7.3 which notably details elements contributing to a renewed policy agenda at regional level.

7.1   Interpretation of the key findings

This section is devoted to the synthesis and interpretation of the main results of the investigation seen in chapters 5 and 6. The findings related to SMEs and KIBS are confronted with results extracted from other studies. The discussion encompasses a selection of empirical surveys each covering some aspects of the present investigation field of the work. In the following, surveys performed on a regional scale are associated with national surveys in order to enhance the perspective in which the findings are interpreted.

Following regional surveys will be integrated in the analysis:

- Herden (1992) related to manufacturing firms in Baden-Württemberg, Bavaria, Switzerland, Liechtenstein and Austria;
- Héraud *et al.* (1993) dealing with manufacturing firms in Alsace and part of Baden-Württemberg.

Additionally, three investigations at national level are considered:

- Brouwer and Kleinknecht (1996) examining manufacturing and service firms in The Netherlands;
- Evangelista and Sirilli (1997) considering manufacturing and services firms in Italy;
- ZEW/FhG ISI (1999) focusing on service firms located in Germany.

The interpretation of the findings focuses on three aspects: (i) the relationships between innovation and performance of SMEs and KIBS; (ii) the existence of a "virtuous circle" linking SMEs and KIBS; and (iii) the influence exerted by territorial determinants on the evolution patterns of the considered firms.

7.1.1   Innovation and evolution: what can be learnt from SMEs and KIBS?

The first conclusion deals with the hypothesis of a **link between the introduction of innovations and the level of economic performance** which was put forward at the beginning of chapter 1. In this respect, innovation has been considered as: (i) a complex learning process, based on interactions; (ii) taking place in SMEs as well as in KIBS; and (iii) an expression of the (positive) evolution of the firm. The results of the empirical investigation support, at least partly, the hypothesis of such a link. Focusing on the economic success of a firm measured in terms of employment growth, a direct link with innovativeness appears only for KIBS, and not for SMEs. More precisely, the introduction of innovations by KIBS has a direct influence on their growth. Innovating SMEs appear to be more successful than non-innovating ones, but this link, in contrast with KIBS, must rather be attributed to indirect effects than to direct ones (*cf.* section 5.3.3. and 6.3.3). In other words, the route from innovation to success seems "shorter" for KIBS than for SMEs. Nevertheless, for both SMEs and KIBS, the results suggest a relation between (employment) growth and innovation. This is coherent with Brouwer and Kleinknecht (1996, p. 38) who, referring to growth not in terms of employment but in terms of sales, assert on the basis of their findings: "*Certainly, the argument that innovators grow more rapidly is plausible. Our database does not allow to investigate the direction of causality.*" The relation between innovation and exportation also highlights the impact of innovation on firms' performance. Even if no highly significant direct influence has been detected, bivariate tests revealed for both SMEs and KIBS a higher propensity of innovating firms to export. In this respect, views expressed in ZEW/ FhG ISI (1999, p. 3) concerning the future of service firms in general can be referred to: "*Innovative service companies export significantly more often than average. These difference will continue to increase: the percentage of exporting companies is expected to grow with innovative service providers, but it will rather diminish as far as the non-innovative ones are concerned.*"

Additionally, considering, in the light of the empirical results, **the respective influences of structural and territorial factors** on the level of performance and evolution capacities of SMEs and KIBS, the strong determination of structural factors, relative to territorial ones, can be established. Both in the case of SMEs and of KIBS, the size of a firm shows a determining impact and one may assert with Brouwer and Kleinknecht (1996, p. 39) that: "*It comes as no surprise that larger firms generally have a higher probability to innovate.*" The examination of sectors of activity revealed a direct influence on the composition of the knowledge base, and to a certain extent, on the evolution patterns of a firm. This is totally consistent with the findings by Herden (1992, pp. 204-206) related to the influence of the type of firms' activity on its innovation-related features. In this respect, the following observation can be made concerning the respective influence of firms' structural and locational factors. As a result of the investigation, it appears that structural factors (sector of activity, size) affect each element of the potential "knowledge ⇒ innovativeness ⇒ success" chain. On the other hand, locational factors (referring to regional environment and to country) show only significant impact (if any) on the composition of the knowledge base of the firm. In fact, it must be remembered that for both samples, the two variables featuring the influence of the location of a firm (*i.e.* in terms of country of location and of regional environment type) on its evolution patterns indicate strongly divergent influences. For both SMEs and KIBS, it has been established that the country of location significantly affects several other variables whereas no direct highly significant impact could be identified depending on the type of regional environment. The aspects related to territorial determinants will be discussed further in section 7.1.3.

Considering highly significant direct influences and summing up the general factors determining the evolution of KIBS and SMEs, the following characterisation can be proposed: (i) **the link between the level of performance of a firm and its innovativeness is stronger in the case of KIBS than in the case of SMEs; and (ii) for both SMEs and KIBS, the level of performance is more "sensitive" to structural than to locational determinants**.

7.1.2 The virtuous circle: between contamination and symbioses

In chapter 2, a second main hypothesis has been expressed concerning the influences of innovation-interactions between SMEs and KIBS on their respective evolution capacities. It has been argued that such interactions may support: (i) a reinforced integration of a firm in its innovation environment; (ii) an improved activation of its internal resources; (iii) as well as of its external ones. In this respect, **the investigation provides strong empirical evidence supporting the hypothesis of a positive impact of interactions with KIBS on SMEs innovativeness**. This impact relates primarily to direct influence (*i.e.* in terms of propensity to innovate), but also encompasses an indirect dimension. In fact, interacting SMEs are characterised by a higher degree of activation of internal and external innovation assets. This is coherent with the views expressed by Herden (1992, p. 196): "*Independently from other determinants (like for example the type of activity, the size, the location, etc.) R&D intensive firms use more technology-oriented external relations than less R&D intensive firms*".[128] In other words, SMEs interacting with KIBS, comparatively to non-interacting SMEs: (i) introduce more innovation; (ii) invest more in R&D; and

---

128 "*Unabhängig von anderen Determinanten (wie beispielsweise der Branchenzugehörigkeit, der Unternehmensgröße, dem Standort des Unternehmens etc.) nutzen FuE-intensive Unternehmen technologieorientierte Außenbeziehungen intensiver als weniger FuE-intensive Unternehmen*" (Herden, 1992, p. 196).

(iii) show a greater propensity to co-operate with ITI.[129] The complementary hypothesis of a positive influence of interactions with SMEs on KIBS evolution capacities can also be supported, but with somehow weaker empirical evidence. **The results of the investigation clearly suggest that proximity-based interactions with SMEs determine the inclination of KIBS to innovate.** However, the only highly significant direct influence that can be generally established concerns the positive impact of interactions with SMEs on KIBS' internal R&D investment.

Questioning the constitution of their knowledge base and their propensity to innovate lead to similar, but not identical, conclusions for KIBS and SMEs. In fact, specific constituents of their respective knowledge bases appear as more determinant for SMEs' innovations than for KIBS'. This is for instance the case if one considers the existence of networking with ITI. KIBS seem to be less "dependent" on such sources than SMEs. This is consistent with the following assertion: "*Knowledge-creating processes in the service industry occur less systematically and depend more on situations than in the manufacturing industry. (...) The knowledge potential of universities, technical colleges and non-university research institutions does not yet play a major role in service industries. These institutions only occasionally offer practice-relevant service-related know-how.*" (ZEW/FhG ISI, 1999, p. 5).[130] Moreover, to consider in addition the results of surveys focusing on the innovation networks of manufacturing firms (*i.e.* Herden, 1992, pp. 145-162 and Héraud *et al.*, 1993, pp. 15-38) help to better assess the "relative weight" of interactions with KIBS for SMEs. In fact, SMEs may be less frequently or less easily in contact with KIBS in the frame of their innovation process than with other information sources (such as customers, suppliers, informal contacts, etc.). Nevertheless, this does not contradict the impact of KIBS on SMEs innovation capacities in the case of successful interaction.

The same is true concerning KIBS: SMEs do not constitute their most frequent or most usual information source.[131] Evangelista and Sirilli (1997), regarding the

---

129 In this respect, *cf.* Antonelli, (1998, pp. 178-179): *"In terms of connectivity and receptivity, knowledge-intensive business services function as holders of proprietary 'quasi-generic' knowledge, implemented by the interactions with customers and the scientific community. Knowledge-intensive business services operate as an active interface between codified knowledge, stored in universities and the research laboratories of other firms, and its tacit counterpart, located within the daily practices of the firm."*

130 *Cf.* also Evangelista and Sirilli (1997, p. 14): *"Non-technological innovations do play an important role in firms' strategies, especially in service firms".*

131 It can be for instance referred to Bilderbeek and Den Hertog (1998 p. 134) considering KIBS in The Netherlands and asserting that: *"The main customers of computer and related IT services are concentrated in the non-manufacturing industries: manufacturing industries as a whole took 19% of total sales. Within the manufacturing sector the high tech industries accounted for just 3% of total sales of computer services. Within the non-manufacturing industries the largest client categories are distinctly business and other services, both accounting for 14% of total sales."*

whole service sector, estimate necessary to reconsider the importance of customers as a source of information for service innovations: *"This is perhaps in contrast with most of the literature on services which emphasises the critical role that the user-producer interactions and customisation play in the innovation process in services"* (Evangelista and Sirilli, 1997, p. 16). Nevertheless, as has been underlined previously for SMEs, this does not contradict the impact of interactions with SMEs on KIBS innovation capacities (in the case of successful interaction) but reveal the necessity to consider those interactions as elements encompassed in a broader frame. These aspects, related notably to the "relative weights and places", will be discussed further in proposing an integrated typology of knowledge interactions implying SMEs and KIBS (*cf.* section 7.2.2).

To summarise, it can be argued that **interacting KIBS and SMEs mutually contribute to their respective innovation capacities, in a similar but not identical way.** This mutual contribution is based on a "core sequence" which can be approximated with three "sub-sequences": (i) the interaction itself; (ii) the resulting knowledge base expansion; and (iii) the ensuing evolution of the firm. These three constituents of the whole phenomenon should not be seen in a linear perspective but as potentially inter-linked in a "knowledge-based loop" thanks to feed-back effects. More precisely, it can be suggested in this respect that the activation of internal resources (like R&D) consists of a stimulation in terms of tacit knowledge which benefits both SMEs and KIBS. In parallel, the activation of external resources based mainly on codified knowledge exchanges (like networking with ITI) is more beneficial to SMEs than to KIBS. In other words, both KIBS and SMEs innovation capacities profit from interactions but on the basis of a different knowledge articulation.

7.1.3 The territorial component: does space really matter?

In chapter 3 it has been suggested, that proximity between interacting SMEs and KIBS as well as locational factors have a significant impact on the propensity of those firms to interact, to innovate and to be economically successful. Consequently to this hypothesis the influence of territorial determinants on the evolution capacity of SMEs and KIBS has been investigated.

The empirical investigation focussed on three kinds of territorial determinants: (i) proximity; (ii) national system of innovation; and (iii) type of regional environment. Concerning **proximity** as a first potential territorial determinant, a contrasted conclusion can be provided: (i) from the point of view of SMEs, no distinctive feature related to proximity-based interactions with KIBS can be identified; (ii) from the point of view of KIBS, on the contrary, proximity-based interactions with SMEs appear as a factor positively influencing KIBS' propensity to innovate. In other words, this suggests that **proximity matters more when information flows from**

SMEs and knowledge is developed by KIBS than when information flows from KIBS and knowledge is developed by SMEs.[132] In this respect it is worth referring to Herden (1992, p. 192) stressing: (i) the adequation between qualification of partners and the needs to be fulfilled that prevail over proximity; (ii) the positive impact of proximity on the efficiency of co-operation; and (iii) the determining influence of a common socio-cultural background between partners.[133] The second territorial determinant investigated relates to the **influence of the national innovation system (NIS)**. The results clearly show that for SMEs their propensity to (i) interact with KIBS; and (ii) to invest in R&D are significantly determined by the country of location. In fact, these two innovation-related indicators are significantly higher for German SMEs than for their French counterparts. Considering KIBS, the NIS clearly determines some part of innovation-related activities (e.g. networking with ITI and R&D intensity) but does not affect the patterns of interaction with SMEs. Consequently, German KIBS, like SMEs, show a greater propensity to innovate than the French ones. Altogether, this provides clear support for the belief that, at least for the samples considered in France and in Germany, **the NIS significantly influences interaction and innovation patterns of firms**. This is fully consistent with for instance Héraud *et al.* (1993, pp. 42-56) who reach similar conclusions in comparing the innovative activities of manufacturing firms in Alsace and in the south of Baden-Württemberg (*i.e.* the Bodensee area): French manufacturing firms are less innovation-oriented than their German counterparts. Finally, the third territorial factor investigated empirically, *i.e.* the **type of regional environment**, revealed no significant influence on the interaction and/or innovation behaviours of SMEs and KIBS. In other words, contrary to expectations, the opposition between core and peripheral regions does not disclose significant consequences in terms of firms' evolution. This partly contradicts the results advanced by Herden (1992, pp. 207-210) who considers that the location of a firm strongly affects the structure of its innovation-oriented relations (particularly with higher education institutions).[134] This **total lack of detectable influence related to the type of regional environment** is put into perspective with the assertion by Brouwer and Kleinknecht (1996, p. 41) considering, for the Netherlands, that: "*A firm's location in a certain region*

---

132 As Antonelli (1998, p. 179) asserts it: "*Firms requiring specific solutions of advice can access the competencies of knowledge-intensive business services, which are able to interface their own localized knowledge with the generic scientific and technological competencies available in the external environment, thereby enhancing their own technological capacity.*" This clearly supports the idea that for knowledge-intensive exchanges proximity between KIBS and their client may be of importance, but for asymmetric reasons.

133 *Cf.* Herden (1992, p. 192): (i) "*Qualität geht vor regionaler Nähe*"; (ii) "*Regionale Nähe ist wichtig für die Effizienz der Zusammenarbeit*"; and (iii) "*Wenn es Probleme gibt, sprechen wir Schwaben eine Sprache*".

134 *Cf.* Herden (1992, p. 207): "*Eine Differenzierung der Unternehmen nach ihrem Standort bestätigt, daß (unabhängig vom unterschiedlichen Besatz einzelner Regionen hinsichtlich der dort vertretenen Branchen, Unternehmensgrößenklassen etc.) deutliche regionale Disparitäten hinsichtlich der Nutzung technologieorientierter Außenbeziehungen bestehen.*"

*has no influence on shares in sales of innovative products (...) However, compared
to firms in more rural regions, firms in urbanized areas of the Netherlands have a
significantly higher probability of selling products 'new to the sector' (...) This is
consistent with the hypothesis that, due to 'information density' and 'spill-over effects' from knowledge centres, urban agglomerations are a better 'breeding place'
for innovation than are rural areas"*. Two possible arguments can be advanced as
an explanation. At first, the regions chosen to test the influence of different environment types do not necessarily cover a variety of situations that are sufficiently
wide in comparison to the (combined) impacts of firms' structural and national-
related determinants. An additional explanation relies on the fact that potentially
most of the firms located in a (marginally) less favoured environment tend to adopt
successful strategies in order to compensate for environmental handicaps.

The findings related to the impact of territorial determinants on SMEs' and KIBS'
evolution patterns can be summed up as follows. Firstly, proximity plays a role in
interaction taking place between SMEs and KIBS. However, as it appears that
proximity-based relations are more determinant when information flows from
SMEs than KIBS than in cases of information flows leading to knowledge constitution by SMEs. Secondly, the evolution patterns of SMEs and of KIBS are strongly
determined by the national innovation system. In this respect, German firms show a
notably greater propensity to interact and to innovate than their French counterparts.
Finally, contrary to proximity and NIS, the type of regional environment cannot be
considered as a significant factor determining the way SMEs and KIBS interact,
innovate and evolve.

7.2   Towards an integrated typology of innovation interactions

The previous discussion related to the key findings of the empirical investigation
allows the consideration of interactions between SMEs and KIBS in a broader
frame. In this respect, a model featuring the spatial dimensions of consultancy expertise is examined in section 7.2.1. Going one step further, the following section
exposes a typology of knowledge interactions implying SMEs and KIBS. Finally,
the last section provides a selection of examples extracted from personal interviews.
This selection aims to illustrate the broad diversity of knowledge-exchange situations encompassing SMEs or KIBS.

7.2.1   KIBS demand and supply response: Wood's model

Wood (1998), in featuring the spatial dimensions of demand-supply interactions
implying consultancy expertise, provides a reference framework which usefully
supplements the examination of interactions between KIBS and SMEs (*cf.* figure

7.1). This reference framework enlarges the scope of the analysis developed up to this point. As Wood (1998, p. 12) describes it, this schematic illustration "*presents the dominant scales of business demand for consultancy expertise, from global to national and regional clients, and the emerging patterns of consultancy supply response*".

Considering the elements advanced by Wood (1998, pp. 12-16), three of them particularly deserve to complete the survey's key findings: (i) the influence of interactions in terms of firms' innovation capacities; (ii) the pattern of transmission of knowledge between firms, notably considering size effects; and (iii) the impact of national and regional contexts on the knowledge-exchanges. The first element, designed with the terms of "business changes", covers the impact of interaction on firms' innovation capacities. In this respect, the concept of "business changes", rejoins the idea of mutual impact or "positive contamination" which has been developed in sections 2.2 and 2.3 for SMEs and KIBS. The second element corresponds to the vision of the knowledge exchanged between firms, notably transmitted from larger to smaller firms. In this context, KIBS play the role of an innovation vector, thanks particularly to the dissemination effect of current best management and technical practices. Finally, the third element, dealing with the variety of situations depending on national and regional contexts, stresses the complementary character of regional-based and national-based interactions.

Figure 7.1: **KIBS demand and supply response from a spatial perspective**

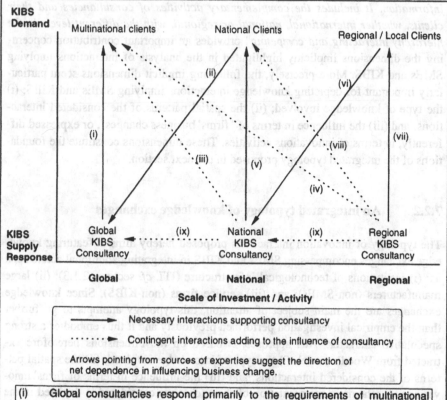

(i) Global consultancies respond primarily to the requirements of multinational clients and agencies.

(ii) Global consultancies increasingly act as conduits of innovative ideas and methodologies between the global and national scales.

(iii) Successful medium-small nationally-based consultancies may develop internationalisation strategies by serving multinational clients seeking specialist expertise or familiarity with home country conditions.

(iv) Successful regionally-based consultancies may also work for multinational clients operating in their regions on a similar basis to (iii), although their growth more often depends on serving national clients (viii).

(v) Nationally-based consultancies, serving private and government clients within that market, provide the predominant volume of consultancy exchanges across a wide variety of expertise.

(vi) Within national systems of consultancy-client interaction, regionally-based clients seeking consultancy support often depend on nationally-based consultancies.

(vii) Regionally-based consultancies originate largely to serve regional clients, and adapt to these needs on the basis of local exchange and innovativeness.

(viii) Successful regionally-based consultancies most often grow by serving national or even international (iv) clients on the basis of specialist skills or knowledge of local conditions.

(ix) Contingent links may exist between international, national and regionally-based consultancies, either directly through subcontracting or networking relationships, or indirectly as a result of client tendering policy.

Adapted from Wood (1998, pp. 13-14, 21)

This vision proposed by Wood (1998, pp. 14-15) of a *"developing global system of expertise exchange, whose significance extends far beyond the simple transfer of information. It includes the complementary activities of consultancies and their clients, whether international, national or regional, with the different levels in the hierarchy interacting and competing"* provides an important contribution concerning the dimensions implicitly highlighted in the analysis of interactions implying SMEs and KIBS. More precisely, the following implicit dimensions seem particularly important for depicting knowledge interactions implying SMEs and KIBS: (i) the type of knowledge involved; (ii) the spatial patterns of the considered interactions; and (iii) the influence in terms of "firms' business changes", or expressed differently, in terms of innovations activities. These dimensions constitute the foundations of the integrated typology proposed in the next section.

7.2.2   An integrated typology of knowledge exchanges

The typology of innovation interactions proposed hereby aims at featuring knowledge exchanges encompassing SMEs and KIBS in integrating additional actors such as: (i) institutions of technological infrastructure (ITI, *cf.* section 2.1.3); (ii) large manufacturers (non-SMEs); and (iii) service firms (non-KIBS). Since knowledge exchanges are the main subject of attention, this typology attempts to go further than the empirical investigation performed previously and it thus embodies a strong speculative character. This approach is based on the dimensions heretofore extracted from Wood's model: (i) the type of knowledge involved; (ii) the spatial patterns of the considered interactions; and (iii) the influence in terms of firms' innovations. A schematic representation of the exchanges considered is depicted in the form of a "wheel of knowledge interactions implying KIBS and SMEs" (*cf.* figure 7.2).

Figure 7.2:   **The wheel of knowledge interactions implying KIBS and SMEs**

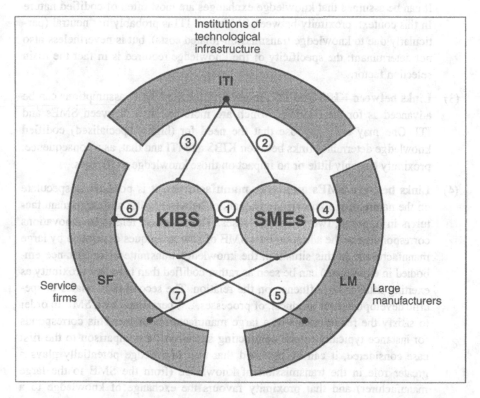

The proposed integrated typology combines - in a speculative way - the key conclusions of the investigation with theoretical elements previously considered.135 As a synthesis, seven main key types of links featuring the knowledge exchanges involving SMEs and SMEs can be schematically depicted:

(1) **Links between SMEs and KIBS.** This relation constituted the core of the performed investigation which provided strong evidence concerning the mutual impact of interaction between SMEs and KIBS on their respective innovation capacities. As a consequence of the empirical results, it has been assumed that: (i) when knowledge flows from SMEs to KIBS the knowledge transmitted is mainly of tacit nature and proximity plays an important role; and (ii) when knowledge flows from KIBS to SMEs, proximity appears to be less important because the transmitted knowledge is mainly of codified nature.

---

135 This concerns notably (additionally to Wood's model of KIBS demand and supply response from a spatial perspective) the concepts exposed in: (i) the "chain-linked" model of innovation (*cf.* section 1.1.2); (ii) the knowledge pyramid (*cf.* section 1.2.3); and (iii) the schematic representation of the possible activities of ITI (*cf.* section 2.1.3).

(2) **Links between SMEs and ITI.** The direct links between SMEs and ITI have as their main function to support and to reinforce SMEs' innovation potential. It can be assumed that knowledge exchanges are most often of codified nature In this context, proximity between SMEs and ITI is probably not neutral (particularly due to knowledge transmission-related costs), but is nevertheless also not determinant: the specificity of the knowledge required is in fact the main selection factor.

(3) **Links between KIBS and ITI.** In general, the same basic assumptions can be advanced as for the relations, which are more common, between SMEs and ITI. One may even suppose that the need for (highly specialised) codified knowledge determines links between KIBS and ITI and that, as a consequence, proximity has only little or no impact on those knowledge exchanges.

(4) **Links between SMEs and large manufacturers.** It is possible to speculate on the nature of the knowledge exchanges between SMEs and large manufacturers in opposing two schematic cases. The first case relates to innovations corresponding to the adoption by a SME of new techniques developed by large manufacturers. In this situation, the knowledge transmitted (for instance embodied in equipment) can be seen as rather codified than tacit and proximity as exerting little or no influence on the relation. The second case relates to specific development (of products, of processes, ...) performed by a SME in order to satisfy the requirements of a large manufacturing client. This corresponds for instance typically to sub-contracting situations. In comparison to the first case considered, it can be assumed that tacit knowledge potentially plays a greater role in the transmission of knowledge (from the SME to the large manufacturer) and that proximity favours the exchange of knowledge to a certain extent.

(5) **Links between KIBS and large manufacturers.** Three typical situations can be used to depict these relationships. The first situation is the adoption by a KIBS of artefacts produced by large manufacturers. The adoption induces associated organisational change (such as for instance in the case of information technologies related equipment). In such a situation, knowledge exchanges can be seen as mainly codified (strongly embodied in the artefacts) and relatively insensible to proximity effects. The second situation relates to knowledge exchanges taking the form of support (for instance managerial consultancy) provided by KIBS to large manufacturers. Such support constitutes a potential source of (internal) innovation for KIBS. It can be interpreted as the application of knowledge which is partly tacit and partly codified . In this case, it is also realistic to consider that it only plays a marginal role. The third situation depicts KIBS resulting from the outsourcing of a specific activity initiated by a large manufacturer. In comparison to the two previous forms of relations, this type of interaction is characterised rather by the circulation of specific tacit knowledge and by an importance given to proximity between the partners.

(6) **Links between KIBS and service firms.** For KIBS service firms represent one of their most important groups of customers. The effects of relations with (non-KIBS) service firms on KIBS' innovation capacities rely mainly on the conception of new services or on the evolution of existing ones (in order to fulfil service firms' emerging or changing needs). These relations can mainly be considered as a process of knowledge codification: (i) which takes place in the client firms; (ii) which is based on KIBS accumulated tacit knowledge; and (iii) for which proximity is rather unimportant.

(7) **Links between SMEs and service firms.** In contrast to the interactions involving KIBS, the knowledge exchanges which take place between SMEs and service firms correspond rather to routine services and thus generate an impact relatively small on SMEs' innovation capacities. The type of knowledge involved and the role of proximity depends strongly on: (i) the degree of standardisation of the activity associating SMEs and service firms; and (ii) the importance of elements like trust for the service relation. Schematically, highly standardised services (typically based on the application of codified knowledge and by which the question of proximity is only related to cost effects) can be compared to more specific services (which imply a greater proportion of tacit knowledge and for which proximity is at least potentially important due to the necessity of a trusting relationship between the SME and the service firm).

7.2.3   Examples

The examples contained in this section are extracted from a set of approximately 40 personal interviews performed in Alsace and Baden during the first semester of 1997. The interviews dealt with SMEs, KIBS, research labs and regional innovation-support organisations identified mainly on the basis of the postal survey.[136] The aim was to gain additional qualitative information supplying the mainly quantitative data collection (presented in chapter 4 and exploited in chapters 5 and 6). In this respect, the following selected examples cannot be expected to have the value of statistical proofs, on the contrary they constitute only illustrations based on stylised facts.[137]

(1) **Links between SMEs and KIBS.** A first example is an Alsatian firm specialised in marketing consultancy for manufacturing firms, particularly small

---

[136] The personal interviews concerned: 8 SMEs, 4 KIBS and 5 ITI in Alsace and 9 SMEs, 6 KIBS and 7 ITI in Baden.

[137] No indication in terms of name or of precise location are given in order to respect the confidentiality of the answers.

ones. At the time of the interview, in addition to a shift of a part of this KIBS' activities in the direction of a reinforced support of clients in terms of financial engineering, the main innovation resided in a new conception of the information system of the firm (*i.e.* definition of standard procedures for client firms data information reception and collection and the establishment of data bases dealing with products and clients). Here the innovation can be interpreted as a process of codification of the knowledge extracted from clients in order: (i) to improve the efficiency of the service relation benefiting the clients; (ii) to favour the constitution of a competitive advantage through an "active memorisation" of information, or in other words, through an expansion of the knowledge base of the consultancy firm. However, the interview revealed some indications which are surprising at first glance: the owner and general manager of this KIBS insisted on the determining role of geographical and cultural proximity between his firm and its clients. Considering the process of codification underlying the innovation examined, it appears that in this case the paradox is only apparent: the knowledge codification process requires an important tacit understanding of the clients situation and of the products specificity and this tacit understanding is reinforced by the proximity underlying the relation.

(2) **Links between SMEs and ITI**. An illustration can be found in the case of a Badian high-tech SME specialised in the electrostatic treatment of surfaces (by machinery equipment, medical instrumentation, buildings, *etc.*). Treatment processes are continuously improved and new ones are regularly tested and introduced. In this respect, co-operation with ITI is common and complements the internal R&D efforts in several ways. However, two precise cases, *i.e.*: (i) the preparation of an ISO 9000 certification; and (ii) the performance of an ecological audit (*Öko-Audit*) are linked intimately with the innovative activity of the firm and reveal the following characteristics of the relation. Firstly, the relation is mainly oriented towards codified knowledge. Secondly, the location of an ITI seems to play no role in its selection as a partner (at least on a national scale): the competencies are determining not the proximity.

(3) **Links between KIBS and ITI**. Some elements are provided by the case of an Alsatian research lab specialised in the fields of automation and electronics. The lab's exchanges are internationally-oriented, mainly towards the car and electronic industries. The lab was formerly in contact with a private legal consultancy office specialised in patenting. Meanwhile, the labs' patenting-related legal support is assumed by one of its industrial co-operation partners: a global player in electronics encompassing one of the biggest patenting departments world-wide. The specific activity related to patenting and dealing with legal expertise and advice constitutes a process which is situated upstream from the introduction of innovation itself by a firm. Nevertheless, and independently from the difficulties with which this relation between an ITI and a KIBS was confronted, two elements can be identified. Firstly, the knowledge link between these two actors encompassed exchanges of both: (i) codified knowl-

edge (*i.e.* the scientific substance to be patented); and (ii) tacit knowledge (*i.e.* the legal know-how related to the patenting procedures). Secondly, from the point of the considered ITI, proximity had little or no relevance for such relations. Another example of a relation between KIBS and ITI can be examined in considering the case of a Badian consultancy firm. This KIBS provides specifically: (i) accounting, tax and legal expertise; and (ii) financial and transmission-related advice to family businesses. During the interview, the manager stressed that this field had until now only been the object of parcelled attention from the academic world. In this respect, pioneer contacts have been developed with two higher education and research institutions in Switzerland. The aim is to encourage and develop a "cross-over" between managerial and human sciences and lessons gained through the firm's practitioners experiences. As such, in terms of knowledge exchanges, this relation can be depicted as a combination of: (i) a transfer from a part of the tacit knowledge gained by the KIBS towards the two selected ITI; and (ii) a knowledge feed-back from the two ITI (supposed to take place in the long run). This feed-back is expected to primarily benefit the consultancy firm, and should consist mainly of codified elements (*i.e.* corresponding to academic standards). Proximity does not seem to affect these relations: the two higher education institutions are localised in Switzerland, the one even in the French-speaking part, so the argument of a cultural (in this case language-based) proximity can also be dismissed.

(4) **Links between SMEs and large manufacturers**. Two partly contradictory examples are interesting to confront here. The first example deals with an Alsatian SME producing shop signs and advertising supports. New production processes (engraving techniques, scanner, etc.) and new organisational (computer-related) methods have been introduced. However, as expressed during the interview, the firm "meets" opportunities to innovate rather than "launches" them. In this respect, its machinery suppliers, *i.e.* larger manufacturers play a determining role in proposing new machines or in adapting production gears. An important aspect in the relation the firm maintains with its suppliers concerns the exchanges of ideas: suggestions formulated by the firm are adopted by suppliers and encompassed in the new equipment proposed. Besides the commercial transactions, the relationship with suppliers leading to the adoption of innovations is constituted of informal exchanges, which can be mainly interpreted in this case as a form of circulation of tacit knowledge. Referring to the interview, those relations are undoubtedly characterised by a strong geographical and cultural proximity as the use of the Alsatian dialect as "working language" and the importance given to common membership to local charitable organisation indicate it. The second example consists of a Badian SME designing and producing printers and data saving systems for the computer industry. The conception of the products is strongly client-specified and as a consequence the firm is innovating continuously in order to meet the requirements of its clients, most of them being large manufacturers. The clients

are spread internationally (USA, Great-Britain, Japan, Mexico) and videoconferencing is commonly used to communicate with them. These exchanges, notably of technical specification, can be seen as consisting mainly of codified knowledge. As one may notice, in this case, the geographical and cultural proximity between partners is definitely not important. Nevertheless, this absence of proximity does not mean that those relations are necessarily less human: in certain circumstances (common test of equipment for instance) national flags are raised in the front of the company's building in the honour of the visiting clients.

(5) **Links between KIBS and large manufacturers.** The case of a small Alsatian KIBS (specialised in engineering) whose field of expertise focuses on the conception of industrial piping and related electronic equipment is worth considering. One of the innovations examined in the interview consists of a combination of product and organisational innovations (partly linked to ISO certification process). More precisely, the combination dealt with the development of a new measurement apparatus encompassing micro-checking tools for one of the biggest European electricity producers. The development of the measurement apparatus was constituted of three stages: (i) the in-house conception of the apparatus; (ii) the external realisation itself (performed by a subcontractor); and (iii) the in-house tests and controls performed on behalf of the contractor. The internal know-how (*i.e.* mainly constituted of tacit knowledge) of the KIBS played a determining role for this product development. This know-how can be seen as embodied in the employees, and thus is particularly difficult to "project" over distance. This may explain the major importance devoted to proximity by the firm: the contractors and sub-contractors need to be easily reachable. As a consequence, most of the partners of the KIBS considered as important (including large manufacturers) were present in a range of 50 km from its location at the time of the interview.

(6) **Links between KIBS and service firms.** Two examples from Baden help to illustrate such links. The first example corresponds to a marketing firm which innovated in introducing organisational changes (and related softwares) allowing it to manage a broader spectrum of customers and of products in parallel. The organisational changes were undertook in order to reinforce the KIBS position related to Tele-Shopping activities and to profit from the emergence of electronic commerce. The partners involved in the changes (in addition to the KIBS developing the specific software) were big wholesale dealers and media-related services. Interpreting those relations in terms of knowledge exchanges, it can be suggested that those exchanges: (i) were mainly based on codified knowledge; (ii) constitute a prerequisite to the application of the KIBS' tacit knowledge within the relation. The fact that some of the partners are located in France and in Great-Britain seems to indicate that proximity did not play a particular role in the considered knowledge exchange. A somehow similar situation can be observed in examining the second example: a KIBS

which assumes (in addition to other consulting activities) a gatekeeper function for a new banking group providing financial investment to innovative "green" start-ups (*i.e.* environmentally aware). This relation can be understood as an addition of the KIBS specific tacit knowledge to the rather codified knowledge-base of the financial institution. In this case also, the proximity between the KIBS and the non-KIBS service firm seems not to constitute a precondition for the relation.

(7) **Links between SMEs and service firms.** The examples chosen deal with two Alsatian SMEs which present partly diverging patterns of relations to service firms. The first SME is a sub-contractor in the textile sector. The interview showed that changes occurring in that firm (mainly production process innovations) are strongly dependant on the acquisition of new equipment. As a consequence, one crucial aspect of the managerial activity deals with investment financing. In this respect, banks constitute determinant partners for the firm. These relationships between the firm and its banks are definitely proximity based. Some attention should be paid particularly to this point since it contrasts the firm's relations with: (i) public technical expertise centres (related to tests and certifications); (ii) equipment suppliers; and (iii) KIBS (managerial consultancy) which t based on proximity. It seems difficult to assess the type of knowledge characterising these relations with banks. Nevertheless, it remains that the importance of trust for such relations suggests hat these knowledge exchanges are rather tacit-oriented than codified-based. The second example relies on the peculiar case of a small Alsatian start-up. This firm, active in the field of printing and of recorded media reproduction, had grown from 1 to 15 employees in less than five years after its creation (at the time of the interview). The introduction of (mainly production process–related) innovations within the firm is intimately linked to the skills and qualifications of the personnel. However, this precise case is somehow peculiar since the main goal of the firm is to support the social insertion of persons in difficulty. In this example, the selected relationship with a (non-KIBS) service firm goes further than the "traditional" acceptation of service firms: it focuses on the connection between the firm and external social workers and services which assist the firm in seeking and training potential new employees. The underlying knowledge relation consists mainly of a selection and a circulation of embodied knowledge, for which proximity plays a determining role.

These few examples, even if they only have an illustrative value, allow a portrayal of the diversity of knowledge exchanges linking SMEs and KIBS to each other and to further firms or institutions. This diversity of situations should be kept in mind when exploring the consequences of the investigation results for policies.

7.3   Implications for policies

Considering the findings of the investigation as well as the more speculative reflections on knowledge exchanges implying SMEs or KIBS, it is possible to gain some
indication for innovation-oriented policies. In this respect, three main aspects are
questioned hereafter: (i) the significance, for policies, of a **broader conception of
innovation**; (ii) the possibilities for policies to integrate the observed **induced support effects**; and (iii) the indications which can be gained for **regional development.**

7.3.1   Innovation: rather a matter of knowledge than of technique?

The "technical bias" affects not only the vision of innovation from a theoretical
point of view but (as stressed in chapter 1) also the content of innovation-related
policies. The growing importance of the intangible (which tends to be undervalued
by the "technicist bias") is true considering service innovations as well as manufacturing industries.[138] In fact, the findings of the investigation underline that
speaking about innovation basically means speaking about knowledge. In this respect, two main implications can be derived: (i) the crucial obligation for innovation
policies not to neglect service firms, and particularly KIBS; and (ii) the relevance
for innovation policies to be oriented rather towards firms' evolution than to be
purely focused on R&D.

Generally speaking, one may assume that innovation policies generally suffer from
a "plant syndrome" (*i.e.* a difficulty to consider other places for innovation than
manufacturing firms) which leads to a strong focus on the manufacturing sector.
However, the results of the investigation clearly show the necessity for innovation
policies to reach a new balance in terms of activities benefiting from innovation
support. In this respect, one of the main challenges for innovation policies probably
deals with a reinforced concern for innovations taking place in the service sector.
This conclusion fully meets the points of view[139] defended by authors like Coffey
and Bailly (1992), Djellal (1993) or Gadrey *et al.* (1993). As a consequence, since
the empirical evidence gained by the investigation support the principle of virtuous
feed-back effects linking SMEs to KIBS, particular attention should be paid to
KIBS by innovation policies.[140]

---

138 As an illustration, the increasing significance of information technologies (IT) for firms as well
as the drastic current impact of organisational innovation, may be perceived as some expression
of the growing "importance of the intangible" for firms.

139 Exposed notably in sections 1.3.1 and 1.3.2.

140 About the policy implications of KIBS for the innovativeness and competitiveness of the German economy for example, one may refer to Strambach (1997) declaring that: "*Die politische
Herausforderung besteht darin, nicht nur den Fortschritt entlang des eingeschlagenen tech-*

Another (and partly complementary) consequence of the "technical bias" denounced previously concerns the strong focus given by innovation policies on (physically or budgetary identifiable) R&D activities instead of concentrating on the expansion of firms' knowledge bases. A broader (or holistic) approach of innovation policy-support would for instance imply tempering any conception favouring mainly: (i) punctual activities (like technology transfer); (ii) exclusive components of the firm (like R&D departments); or (iii) restricted specific behaviours (like formalised R&D investments at the exclusion of other expenses). Consequently, innovation policies should adopt procedures more oriented towards learning effects and knowledge transformation. In this respect, KIBS may play a particular "gate-keeping" role since, as Antonelli (1998, p. 181) asserted: "[i]n Europe, the historical antagonism, or at least the lack of established communication and cooperation, between industry and the academy may well be bridged by knowledge firms acting as connectors and mediators in the exchange of technological information". More generally, it would be worthwhile to implement such procedures on the basis of principles like the concept of induced support discussed next.

7.3.2 The concept of induced support

The results of the investigation suggest reconsidering the respective impact of ITI and of KIBS on firms' innovativeness. This re-interpretation derives from the basic assumption that firms' needs in terms of knowledge-intensive support are probably underestimated whereas firms' needs related to technical or scientific information are overestimated. As a logical consequence, advocating in favour of an alternative vision of innovation interactions, it seems worthwhile to hypothesise a new vision of the "triangle": (i) scientific production; (ii) manufacturing activities; (iii) knowledge-intensive service activities. In this respect, one may for instance refer to Tomlinson (1997, p. 16). who analyses the contribution of services to the manufacturing industry as follows: "*Knowledge based business services inputs appear to be highly significant whether we look evidence that these service inputs are less important for manufacturing output than at their influence within the services or manufacturing sectors. In fact there is no fixed capital. There is therefore a convincing argument that manufacturing sectors in post-industrial societies rely on these services. Rather than the manufacturing base being eroded away, the development of specialised services may in fact be beneficial to manufacturing. The manufacturing versus service dichotomy is then unhelpful. The knowledge based service sector is an inte-*

---

nologie- und industrieorientierten Pfades zu stimulieren, was sicherlich weiterhin notwendig ist und mit einem vielfältigen Maßnahmenbündel bereits getan wird, sondern auch neue Handlungsfelder im Bereich wissensintensiver Dienstleistungen zu erschließen. (...) Um Anpassungsfähigkeit und Innovationsprozesse zu fördern, sind keine Branchenstrategien, sondern Maßnahmen zur Unterstützung der Interaktion zwischen Angebot und Nachfrage erforderlich." (Strambach, 1997, p. 240).

*gral part of the economic sector rather than an 'unproductive' or 'parasitic lag-
gard"*.

Basically, the traditional linear conception considers that the "scientific system"
generates the inputs leading firms to innovations. These innovations are supposed to
take place mainly in manufacturing firms, service firms being reduced in this vision
to a peripheral role or to an accompaniment function. A more realistic vision, allowing the existence of an "induced support", would consider that innovation, most
often, corresponds rather to a "demand pulled" than to a "science pushed" phenomenon. Consequently, there is no serious reason to believe that "scientific inputs"
are *a priori* more important for manufacturing than for service activities. On the
contrary, the development of information technologies (IT) and the progresses ac-
complished by social sciences plead in favour of a growing impact of the "scientific
system" on service firms, especially on knowledge-intensive ones. This vision constitutes a strong incentive to modify the somewhat old-fashioned way of thinking
about firms' innovation activities and science.[141]

Looking at the policy implications of the principle of "induced support", the consequences of the virtuous circle linking SMEs and KIBS can be expressed in the simplest terms: "helping the ones" will contribute to "help the others".[142] Basically, the
"induced support" approach suggests that innovation policies should primarily focus
on the "cognitive value" of firms, *i.e.* on their strategies in terms of knowledge expansion and composition. In this respect, the principle of an "induced support" is
intimately linked to KIBS since, as previously exposed KIBS: (i) have a particular
impact on the evolution capacities of SMEs; (ii) play the role of an innovation vector, besides other actors such as ITI; (iii) in contrast to ITI, are not only innovation
vector but also (potential) innovators. This KIBS-related "induced support" effect,
which is relevant from a general perspective, is even more critical if one considers
regional policies. As Wood (1998, p. 12) put it: "*Consultancies promote change and
innovation especially when transmitting experience from larger to smaller clients or
from specialist to less specialist clients (for example within a particular industrial
or regional innovation system). The nature of these demand-supply interactions
varies widely between different scales of activity and between types of consultancy
expertise and in different regional and national economic environments.*"

---

141 Indirectly, this point is supported also by Gadrey (1994, p. 38) asserting that "(...) *les échanges
d'informations complexes et de connaissances sont donc de plus en plus l'enjeu de ces relations*
[service relations], *qui visent alors la coproduction de la maîtrise intellectuelle de la complexité
et de la réduction de l'incertitude dans les décisions*".

142 To this aspect, and more precisely on the impact of particular categories of KIBS on manufacturing firms, it can be referred to the evaluation exercise of the "consultancy initiative" performed on behalf of the British Department of Trade and Industry (DTI) by Segal Quince
Wicksteed Limited (1991).

7.3.3 Some elements contributing to a renewed regional policy agenda

The findings of the investigation question the concept of territory, and consequently provide elements feeding the reflection oriented towards regional development policies. It is possible to give a frame to the discussion in considering the heterodox approach (or "holy trinity" of regional economics, *cf.* figure 7.3) proposed by Storper (1997). In this approach, the *guiding metaphor* is no more economic systems as machines but the economy as sets of relations: "*Technology involves not just the tension between scale and variety; but that between the codifiability or noncodifiability of knowledge; its substantive domain is learning and* becoming, *not just diffusion and deployment.* (...) *Territorial economics are not only created, in a globalizing world economy, by proximity in input-output relations, but more so by proximity in the untraded or relational dimensions of organizations and technologies. Their principal assets – because scarce and slow to create and imitate – are no longer material, but relational.*" (Storper, 1997, p. 28). However, some clear discrepancies appear when one tries to establish a convergence between the results of the investigation, the vision proposed by Storper (1997) and the realities of regional development policies. In fact, institutions in charge of such policies refer spontaneously to territories in terms of administrative (or "political") units. Nevertheless, such units do not obligatorily correspond to the patterns adopted by firms, especially in terms of knowledge flows and innovation-related strategies. In this respect, bearing in mind the complexity of the concept of territory, two issues faced by regional innovation policies [143] are examined hereafter: (i) the absence of a "territorial fatality"; and (ii) the necessity to consider regions as **open systems**.

---

[143] *Cf.* for instance Morgan (1997) focusing particularly on the new generation of EU regional policy measures for the aspects related to the "learning region"

Figure 7.3:   **The "holy trinity" of regional economics according to Storper**

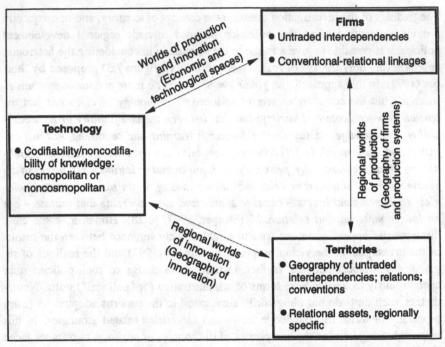

Adapted from Storper (1997, p. 27 and p. 42)

To refute territorial fatality means to suppose that, even if some firms benefit from positive territorial determinants[144], other firms, located in less favourable environments, tend to adopt strategies enabling them to compensate for their "locational handicaps". Examining the results of the investigation dealing with the influence of territorial factors on the evolution patterns of KIBS and SMEs , one may assume, in a metaphoric way, that: (i) those firms are more "organic" than "mineral" (*i.e.* they are to a certain extent sensitive to their environment); and (ii) those firms, being "organic", are more "animal" than "vegetal" (*i.e.* they adapt rapidly to environmental changes). As a consequence, considering firms' capacity to adopt successful strategies compensating for their environmental weaknesses on the one hand and

---

144 For instance, concerning the influence of spatial determinants on innovative behaviours of firms, Camagni and Capello (1997, p. 20) assert that: "*The spatial environment provides in fact some key elements, some district economies, which can support local innovativeness: the level of integration among firms (cooperation capability), also of different size (systematisation capability) and integration with the urban services (capacity to exploit urbanisation externalities) (...) however, the existence of pure district economies does not automatically lead to innovation (...) moreover, in some extremely turbulent and innovative economic phases, these local dynamic elements require to be complemented by cooperative behaviour mechanisms, like cooperation agreements with firms outside the area.*"

the potentialities offered by the concept of induced support discussed above, on the other hand, one may claim that **even if the territory exerts an influence, there is no territorial fatality.**[145] In this respect, bearing in mind the virtuous evolutionary circle linking SMEs and KIBS demonstrated by the investigation, it seems particularly relevant for policy recommendations to examine suggestions such as the ones expressed for instance by Marshall (1982). Having analysed the manufacturing industry's demand for business services in three British city regions, this author advocated for a policy-supported-expansion of service activities as a successful strategy for regional development and asserted that: *"Such a strategy should be more easily pursued by government than the traditional forms of regional policy in manufacturing industry, because financial encouragement needs to be less substantial in service industries (...)"* (*cf.* Marshall, 1982, p. 1537).

The final point relates to the inescapable "openness" of regional systems. Regional policies, implemented notably in less-favoured regions, often try to overcompensate for environmental weaknesses; this may lead, in extreme cases, to the constitution of "cathedrals in the desert". Such a tendency reveals the inclination to consider territories as close systems and to systematically favour relations taking place between actors from the (territorial) system. According to the results of the investigation, it can be suggested that to concentrate on firms' learning processes would be more efficient for regional policies than to try to overdevelop some regional innovation-related infrastructure. Such efforts, dealing with firms' learning processes, should particularly concern innovation-oriented interactions, and should not be restricted to actors located in the region exclusively. Innovations-interactions (between firms as well as between firms and their environment) are based on knowledge exchanges which are not limited to the administrative *a priori* unit defined by border which constitute a "political" region.[146] Thus, the development of the knowledge-based economy reveals the hidden dimensions of proximity. In this respect, Nielsen (1996) can be referred to as an example who stresses a paradox affecting, in his opinion, particularly small firms: *"On the one hand, it is imperative for even small firms to have access to knowledge which is at least at the same level as that of the competitors. The competition is international, hence the knowledge*

---

145 In this respect, one may refer to Capron and Debande (1997, p. 692) concluding on recent researches related to the role of the manufacturing base in the development of services (mainly in industrial declining regions in Belgium) that: *"The results obtained in [the] analysis emphasize the need for an integrated regional policy whose objectives should be, on the one hand, to stimulate new manufacturing activities in high technology areas able to strengthen the regional manufacturing base and, on the other hand, to promote the creation of complementary high level business service activities."*

146 This assertion is however subject to discussion, *cf.* for instance Ohmae (1993, p. 78) assuming that: *"The boundaries of the region state are not imposed by political fiat. They are drawn by the deft but invisible hand of the global market for goods and services. They follow, rather than precede, real flows of human activity, creating nothing new but ratifying existing patterns manifest in countless individual decisions."*

has to be international state-of-the-art. (...) On the other hand, there is - especially among small firms - a strong attachment to the local environment. Adequacy of information about available technology decreases with distance, and constraints against working with an institute increase with distance" (Nielsen, 1996, p. 247). In fact, focussing on knowledge exchanges involving SMEs, KIBS and other actors such as ITI, large enterprises and (non-KIBS) service firms (*cf.* section 7.2.2), **the necessity for regional innovation policies to think in terms of "open system"** implies the adoption of a broader vision of regional innovation capacities.[147] In this respect, one of the main consequences would be learning to think about regions in terms of "areas with fluent borders" or of "knowledge-territories" (not necessarily corresponding to administrative units).

Conclusion

This chapter provides a general interpretation of the investigation's key findings in confronting the statistical results obtained with the empirical evidence exposed in other analyses. Three elements constitute the core of this discussion: (i) the question of a link between the innovativeness of SMEs and KIBS and their economic performance; (ii) the virtuous circle linking the evolution capacities of SMEs and KIBS; and (iii) the influence of spatial factors, notably the influence of proximity which is related to the type of knowledge exchanged. Succeeding this reflection, an enhanced vision of knowledge exchanges encompassing SMEs and KIBS is proposed by taking into consideration additional actors such as ITI, large manufacturers and (non-KIBS) service firms. This attempt (entitled the "wheel of knowledge interactions") corresponds to a speculative display of knowledge exchanges in the form of an integrated typology. Considering the implications for policies, the main contribution of the investigation is probably to put knowledge at the core of the innovation problem. In this respect, it seems worth paying a particular attention to the virtuous circle linking SMEs and KIBS as a policy tool, allowing notably induced innovation support. Last but not least, the completed analysis offers further possibilities to regional policies and thus contributes to a renewed agenda for regional development in Europe.

---

147 These aspects are emphasised by Koschatzky and Muller (1997, p. 23) who assert that: "(...) *our results recommend a more broader view in regional innovation policy. Especially in advanced economies with close infrastructural linkages, measures directed towards single, politically defined regions should be subject of a better coordination for the generation of interregional synergies. Cross-regional activities would increase the impact of regionally oriented measures and by this provide a stronger support to innovation and competitiveness of local and regional firms.*"

General conclusion

Considering the results of the investigation, it can be assumed that the main questions underlying the analysis have been answered. Referring notably to the different hypotheses expressed, it can at first be considered that innovations, resulting from complex learning processes based on interactions, constitute an expression of the evolutionary capacities of SMEs and KIBS. In this respect, the relationship between the introduction of innovations and SMEs' and KIBS' level of economic performance has been questioned. A direct link between innovativeness and employment growth could be established at least for the case of the KIBS examined. In other words, the route from innovation to success seems "shorter" for KIBS than for SMEs. Secondly, examining the mutual contribution (or "co-evolution") of SMEs and KIBS in terms of innovation capacities, the investigation provided strong empirical evidence supporting the positive impact of interactions with KIBS on SMEs' innovativeness. Additionally, the results clearly suggest that proximity-based interactions with SMEs notably determine the inclination of KIBS to innovate. It can be argued that interacting KIBS and SMEs mutually contribute to their respective innovation capacities, in a similar but not identical way. These findings, corresponding to the "virtuous circle hypothesis", indicate that interactions between SMEs and KIBS benefit those firms due to their reinforced integration in their innovation environment, as well as due to an improved activation of their internal and external innovation resources. The third main category of results relates to the spatial factors determining SMEs' and KIBS' evolution patterns. Notably, it appeared that proximity matters more when information flows from SMEs and knowledge is developed by KIBS than when information flows from KIBS and knowledge is developed by SMEs. Concerning location-related determinants, the following elements were ascertained. On the one hand, evolution patterns of SMEs and KIBS are strongly determined by their national innovation system: German firms show a greater propensity to interact and to innovate than their French counterparts. On the other hand, location according to the type of regional environment could not be identified as a factor explaining significant differences in the way SMEs and KIBS interact, innovate and, more generally, evolve.

The potential contribution of the investigation in terms of consequences for policy-making relies mainly on three key implications. At first, it is assumed that innovation is rather a question of knowledge than of technique. Subsequently, this suggests a "new balance" in terms of innovation policies. For instance, the development of support programmes dealing less with "pure R&D" and more with "knowledge expansion" (*i.e.* promotion of firms' cognitive capacities) could be recommended. As a complement to this readjusted balance, it seem worthwhile to re-orient, at least partly, policy innovation-support efforts from the manufacturing to the service industry. Secondly, the confirmation of the hypothesis of a virtuous circle (or "co-evolution") making the link between SMEs' and KIBS' innovation capacities discloses new opportunities for innovation policies. The proposed concept of induced support means that to invest in KIBS' innovation capacities would have a positive impact on SMEs (at least on the ones interacting with KIBS) and since corresponds to a kind of "leverage effect". In terms of support for innovation potential, this would allow KIBS to undertake a role complementary to the one notably played by the technology-transfer infrastructure. Finally, the investigation provides elements contributing to a renewed regional policy agenda. Two elements can be particularly mentioned in this respect: the absence of territorial fatality (since firms are able to adopt successful strategies allowing them to compensate environmental weaknesses) and the necessity to consider regions as open systems (since firms notably do not limit their networking activities according to administrative borders).

In conclusion, it is possible to give some indications for further research. First, an important issue deals with the analysis, determination and assessment of the observed phenomena. It appears clearly that the measurement of firms' innovation activities and more generally, of the firms' evolution patterns are still suffering from strong deficits. These deficits relate to the difficulties inherent to the computation of concepts like "knowledge-related interactions", as well as to the spatial factors encompassed by intangible phenomena. Furthermore, although the concept of innovation is presently related to a solid research tradition, it constitutes a perpetually renewed source of difficulties since the forms of innovation themselves are changing.[148] Such difficulties, which hinder the overall understanding of innovation, are particularly accentuated when service activities are considered.[149] One of the main questions to be addressed on this point is to establish if such difficulties are generated by the specific character of service activities[150] or if they are primarily inherent to inadequate measurement tools.[151] An additional interrogation relates to the

---

148 As stressed by Callon, Larédo and Rabeharisoa (1997, p. 36): *"La triple déconnexion (...): déconnexion entre innovation et productivité, déconnexion entre innovation et recherche, déconnexion entre innovation et technologies, rend problématique le travail d'observation et d'analyse statistique".*

149 *Cf.* Strambach (1998, p. 3) who stresses that: *"It is also evident that, because of the heterogeneity of the service branches, standardised surveys do not do justice to the complexity and the specific characteristics of the innovation processes of knowledge intensive business service firms."* More generally, concerning the difficulties linked to the design and measurement of innovation see Voss *et al.* (1992).

150 In this respect, *cf.* for instance Hipp (1998, p. 11): *"Knowledge based on personal experience is very important for the creation of new knowledge for KIBS. A major difference between KIBS and non-KIBS is reflected by the degree of internal research and development. KIBS conduct R&D more often than non-KIBS, and they do so more continuously. Furthermore, KIBS are characterized by an intensive externalisation process. KIBS also use explicit external sources of knowledge more often (...) The conversion from explicit to tacit knowledge through internal codified knowledge is significantly more common for KIBS than it is for other innovators."*

151 As Evangelista and Savona (1998, p. 4) affirm it: *"The experience accumulated over the last years in measuring innovation in the manufacturing through innovation surveys represents in this respect a useful starting point for measuring innovation in services. (...) However, the question of whether and to what extent the methodological and conceptual framework devel-*

relevance of the distinction usually operated between manufacturing and service activities when knowledge-related activities and evolution patterns are considered. Beyond the statistical measurement difficulties, it seems quite obvious that new concepts are needed, particularly a vision of the firm respecting its inherent complexity. In fact, a firm is at the same time a social system, a changing and learning entity, a result of history, and of technology. This indicates more generally the second main element for forthcoming investigations. In this respect, a better understanding of the nature of knowledge, and notably of firms' knowledge base, constitutes an inescapable necessity. In fact, this requests a shift from the economics of technical change to the evolutionary economics of the firm.[152] Finally, in parallel to this new vision of the firm, this displays a major issue: a new conception of the economics of science and more generally, of the economics of knowledge.[153] An important research issue deals with the understanding and evaluation of intellectual assets.[154] Furthermore, it seems interesting to try to interpret innovations in terms of convergence between knowledge creation and added value.[155] In this respect, a related issue concerns the knowledge economy as a whole and the specific impact of new information and communication technologies, since they alter the conditions of access, retrieval, processing and communication of all types of information.[156] Finally, all of these elements strongly suggest that the announced "knowledge revolution" may lead to a shift from economics of knowledge into "economics of hope".[157]

---

oped with reference to manufacturing activities can be used for analysing and measuring innovation activities in the service sector remains a crucial one."

[152] This pleads particularly in favour of the development of alternative indicators to traditional or established R&D-based ones as for instance Brouwer and Kleinknecht (1996 and 1997) strongly suggest it ("measuring the unmeasurable").

[153] In this respect see notably Dasgupta and David (1994) and Callon and Foray (1997).

[154] To this end, Miller (1996, p. 69) proposes alternative forms of assessment of knowledge based on human capital accounting: "(...) some measures show that total industrial intangible investment had passed physical investment in Germany, Sweden and the United Kingdom by 1987. Other evidence of the changing nature of investment is the increased complementarity between physical and intangible investment as well as the high technology content in both kinds of investments."

[155] As an example, it can be referred to the pioneer works of Tomlinson (1997) who demonstrates the importance of knowledge generated by KIBS through the resulting value added by client firms.

[156] Cf. for instance Antonelli (1998, p. 177) asserting that: "Specifically, they [new information and communication technologies] increase the separability, tradability and transportability of information, thus favouring the commercial opportunities of knowledge-intensive business service firms."

[157] Concerning the concept of "knowledge revolution", one may refer to Chichilnisky (1998, pp. 39-40): "We are undergoing a social and economic revolution which matches the impact of the agricultural and industrial revolutions. This is a 'knowledge revolution' driven by knowledge and by the technologies for processing and communicating it. (...) This revolution brings the hope of a society in which economic progress need not mean increasingly extensive use of earth's resources."

References

Acs, Z.J./Audretsch, D.B. (1990): Innovation and Small Firms. The MIT Press, Cambridge, London.

Alter, N. (1990): La gestion du désordre en entreprise. Éditions L'Harmattan, Paris.

Amendola, M./Gaffard, J.L. (1988): La dynamique économique de l'innovation. Economica, Paris.

Ancori, B. (1983): Communication, information et pouvoir. In: Lichnerowicz, A./Perroux, F./Gadoffre, G. (Eds.): Information et communication. Maloine. pp. 59-84.

Ancori, B./Héraud, J.-A. (1987): Fonctionnalisation de l'activité et information. Document BETA, Université Louis Pasteur, Strasbourg.

Anselin, L./Varga, A./Acs, Z.J. (1997): Entrepreneurship, Geographic Spillovers and University Research: a Spatial Econometric Approach. Paper presented at the International Conference on Innovation and Performance of SMEs. 17th March 1997, Cambridge.

Antonelli, C. (1998): Localized Technological Change, New Information Technology and the Knowledge-based Economy: The European Evidence. In: Journal of Evolutionary Economics. 8. pp. 177-198.

Arrow, K. (1962): The Economic Implication of Learning by Doing. In: Review of Economic Studies. Vol. 29, n. 80. pp. 155-173.

Attali, J. (1998): Dictionnaire du XXIe siècle. Fayard, Paris.

Autès, M. (1995): Le sens du territoire. In: Recherches et Prévisions. 39. pp. 57-71.

Aydalot, P. (Ed.) (1986): Milieux innovateurs en Europe. GREMI, Paris.

Aydalot, P./Keeble, D. (1988) (Eds.): High Technology Industry and Innovative Environment: the European Experience. Routledge, London, New York.

Becattini, G. (1979): Dal settore industriale al distretto industriale. Alcune considerazioni sull'unita di indagine dell'economica industriale. In: Rivista di Economica e Politica Industriale. 1. pp. 7-21.

Becattini, G. (1992): Le district marshallien: une notion socio-économique. In: Benko, G./Lipietz, A. (1992) (Eds.): Les régions qui gagnent - Districts et réseaux: les nouveaux paradigmes de la géographie économique. Presses Universitaires de France, Paris. pp. 35-55.

Bellet, M./Colletis, G./Lecoq, B./Lung, Y./Pecqueur, B./Rallet, A./Torre, A. (1992): Et pourtant ça marche! Quelques réflexions sur l'analyse du concept de proximité. In: Revue d'économie industrielle. 61. pp. 111-128.

Benko, G./Lipietz, A. (1992): Le nouveau débat régional: positions. In: Benko, G./Lipietz, A. (Eds.) (1992): Les régions qui gagnent - Districts et réseaux: les nouveaux paradigmes de la géographie économique. Presses Universitaires de France, Paris. pp. 13-32.

Benko, G./Lipietz, A. (Eds.) (1992): Les régions qui gagnent - Districts et réseaux: les nouveaux paradigmes de la géographie économique. Presses Universitaires de France, Paris.

Benzécri, J.-P. (1992): Correspondence Analysis Handbook. Marcel Dekker, New York.

Bilderbeek, R./Den Hertog, P. (1998): Technology-based Knowledge-intensive Business Services in the Netherlands: Their Significance as a Driving Force Behind Knowledge-driven Innovation. In: Vierteljahreshefte zur Wirtschaftsforschung. 67-2. pp. 126-138.

Braczyk, H.J./Cooke, P./Heidenreich (1998) (Eds.): Regional Innovation Systems - The Role of Governance in a Globalized World. UCL Press, London.

Brouwer, E./Kleinknecht, A. (1996): Alternative Innovation Indicators and Determinants of Innovation. Report to the European Commission. Office for Official Publications of the European Communities, Luxembourg.

Brouwer, E./Kleinknecht, A. (1997): Measuring the Unmeasurable: a Country's Non-R&D expenditure on Product and Service Innovation. In: Research Policy. 25. pp. 1235-1242.

Bughin, J./Jacques, J. M. (1994): Managerial Efficiency and the Schumpeterian Link Between Size, Market Structure and Innovation Revisited. In: Research Policy. 23. pp. 653-659.

Callon, M./Foray, D. (1997): Nouvelle économie de la science ou socio-économie de la recherche scientifique? In: Revue d'économie industrielle. 79. pp. 13-32.

Callon, M./Larédo, P./Rabeharisoa, V. (1997): Que signifie "innover" dans les services - une triple rupture avec le modèle de l'innovation industrielle. In: La Recherche. 295. pp. 34-36.

Camagni, R. (1991): Local "Milieu", Uncertainty and Innovation Networks: Towards a New Dynamic Theory of Economic Space. In: CAMAGNI, R. (Ed.) (1991): Innovation Networks: Spatial perspectives. Belhaven Press, London, New York. pp. 121-144.

Camagni, R. (Ed.) (1991): Innovation Networks: Spatial perspectives. Belhaven Press, London, New York.

Camagni, R./Capello, R. (1997): Innovation and Performance of SMEs in Italy: the Relevance of Spatial Aspects. Paper presented at the International Conference on Innovation and Performance of SMEs, 17 March 1997, Cambridge.

Capaldo, G./Corti, E./Greco, O. (1997): A Coordinated Network of Different Actors to Offer Innovation Services to Develop Local SMEs Inside Areas with a Delay of Development. Paper presented at the ERSA Conference, August 26-29 1997, Rome.

Capron, H./Debande, O. (1997): The Role of the Manufacturing Base in the Development of Private and Public Services. In: Regional Studies. 31/7. pp. 681-693.

Châtelet, G. (1998): Vivre et penser comme des porcs. De l'incitation à l'envie et à l'ennui dans les démocraties-marchés. Exils Éditeur, Paris.

Chichilnisky, G. (1998): The Knowledge Revolution. In: The Journal of International Trade & Economic Development. 7/1. pp. 39-54.

Clark, N. (1986): Introduction: Economic Analysis and Technological Change: a Review of Some Recent Developments. In: MACLEOD, R. (Ed.) (1986): Technology and the Human Prospect. Pinter Publishers, London. pp. 1-27.

Coase, R.H. (1937): The Nature of the Firm. In: Economica. 4. pp. 386-405.

Cobbenhagen, J./Den Hertog, F. (1994): Successful Innovating Firms. What Differentiates the Front Runners? Paper presented at the 1994 Spring Meeting of the Six Countries Programme "Innovation - Applying New Ideas for Profit", London, 25 & 26 May, 1994.

Coffey, W.J./Bailly, A.S. (1992): Producer Services and Systems of Flexible Production. In: Urban Studies. 29/6. pp. 857-868.

Coffey, W.J./Polèse, M. (1992): Producer Services and Regional Development: a Policy-Oriented Perspective. In: Papers of the Regional Science Association. 67. pp. 13-28.

Cohen, W./Levinthal, D. (1989): Innovation and Learning, the Two Faces of R&D. In: Economic Journal. 99.pp. 569-596.

Cohendet, P. (1994): Relations de service et transfert de technologie. In: De Bant, J./Gadrey, J. (Eds.) (1994): Relations de service, marché de services. Editions CNRS. pp. 201-213.

Cooke, P. (1998): Origins of the concept. In: Braczyk, H.J./Cooke, P./Heidenreich (1998) (Eds.): Regional Innovation Systems - The Role of Governance in a Globalized World. UCL Press, London. pp. 2-25.

Cooke, P./Boekholt, P./Schall, N./Schienstock, G. (1996): Regional Innovation Systems: Concepts, Analysis and Typology. Paper presented during the EU-RESTPOR Conference "Global Comparison of Regional RTD and Innovation Strategies for Development and Cohesion". Brussels, 19-21 September.

Curien, N. (Ed.) (1992): Économie et Management des entreprises de réseau. Economica, Paris.

Cyert, R./March, J. (1963): A Behavioural Theory of the Firm. Prentice Hall, Englewood Cliffs.

Daniels, P.W./Bryson, J.R. (1998): A New Language for a New Economic Geography? Revisiting the Manufacturing-Service Divide. Paper presented at the VIII Annual Conference of the European Research Network on Services and Space (RESER), Berlin, 8-10 October 1998.

Dasgupta, P./David, P.A. (1994): Toward a New Economics of Science. In: Research Policy. 23. pp. 487-521.

Davelaar, E.J. (1991): Regional Economic Analysis of Innovation and Incubation. Avebury, Aldershot.

David, P.A./Foray, D. (1995): Accessing and Expanding the Science and Technology Knowledge Base. In: Special Issue on Innovation and Standards, STI (Science Technology Industry) Review No. 16. Organisation for Economic Co-operation and Development (OECD), Paris, pp. 13-68.

Davis, G.B. (1974): Management Information System: Conceptual Foundation, Structure and Development. Mac Graw Hill.

De Bandt, J./Gadrey, J. (Eds.) (1994): Relations de service, marchés de services. CNRS Editions, Paris.

De Bresson, C./Amesse, F. (1991): Networks of Innovators: A Review and Introduction to the Issue. In: Research Policy. 20. pp. 363-379.

Djellal, F. (1993): Les firmes de conseil en technologie de l'information comme agents d'un paradigme socio-technique: analyse de leur organisation fonctionnelle et spatiale. Thèse de Doctorat en Sciences Economiques, Université de Lille.

Dosi, G. (1982): Technological Paradigms and Technological Trajectories: A Suggested Interpretation of the Determinants and Directions of Technical Change. In: Research Policy. 11/3. pp. 147-162.

Dosi, G./Freeman, C./Nelson, R./Silverberg, G./Soete, L. (Eds.) (1988): Technical Change and Economic Theory. Pinter Publishers, London.

Dosi, G./Marengo, L./Fagiolo, G. (1996): Learning in Evolutionary Environments. Working Paper, Department of Economics, University of Trento, Trento.

Edquist, C. (Ed.) (1997): Systems of Innovation. Technologies, Institutions and Organizations. Pinter Publishers, London.

Eliasson, G. (1996): Firm Objectives, Controls and Organization - The Use of Information and the Transfer of Knowledge within the Firm. Kluwer Academic Publishers, Dordrecht.

Ergas, H. (1986): Does Technology Policy Matter? OECD, Paris.

EUROSTAT (1993): NACE Rev. 1. Eurostat, Unit B6 - Nomenclatures, Luxembourg.

Evangelista, R./Savona, M. (1998): Patterns of Innovation in Services. The Results of the Italian Innovation Survey. Paper presented at the VIII Annual Conference of the European Research Network on Services and Space (RESER), Berlin, 8-10 October 1998.

Evangelista, R./Sirili, G. (1997): Innovation in Services and Manufacturing: Results from the Italian Surveys. Working Paper No. 73. ESRC Centre for Business Research, Cambridge.

Ewers, H.J./Allesch, J. (Eds.) (1990): Innovation and Regional Development. Strategies, Instruments and Policy Coordination. Walter de Gruyter, Berlin.

François, J.-P./Goux, D./Guellec, D./Kabla, I./Templé, P. (1996): Le développement d'un outil pour mesurer les compétences: l'enquête compétence pour innover. Paper presented at the OECD Conference on New S&T Indicators for a Knowledge-based Economy. Paris, 19th –21st June 1996.

Freeman, C. (1982): The Economics of Industrial Innovation. Pinter Publishers, London.

Freeman, C. (1987): Technology Policy and Economic Performance: Lessons from Japan. Pinter Publishers, London.

Freeman, C./Clark, J./Soete, L. (1982): Unemployment and Technical Innovation: A Study of Long Waves and Economic Development. Pinter Publishers, London.

Freeman, C./Soete, L. (Eds.) (1992): New Explorations in the Economics of Technical Change. Pinter Publishers, London.

Gadrey, J. (1994): Les relations de service dans le secteur marchand. In: De Bandt, J./Gadrey, J. (Eds.) (1994): Relations de service, marchés de services. CNRS Editions. pp. 23-41.

Gadrey, J./Gallouj, F./Lhuillery, S./Weinstein, O. (1993): La R-D et l'innovation dans les services. Rapport, Ministère de l'Enseignement Supérieur et de la Recherche.

Gadrey, J./Gallouj, F./Weinstein, O. (1995): New Modes of Innovation. How Services Benefit Industry. In: International Journal of Service Industry Management. 6/3. pp. 4-16.

Gallouj, F. (1992): Economie de l'innovation dans les services: au-delà des approches industrialistes. Thèse de Doctorat en Sciences Economiques, Université de Lille, Lille.

Gallouj, F. (1994): Economie de l'innovation dans les services. Editions L'Harmattan, Paris.

Gassmann, O. (1997): Internationales F&E-Management. Oldenbourg Verlag, München, Wien.

Grabher, G. (Ed.) (1993): The Embedded Firm. On the Socioeconomics of Industrial Networks. Routledge, London, New York.

GREMI (1989): Milieux innovateurs et réseaux transnationaux: vers une nouvelle théorie du développement spatial. Actes du colloque GREMI-EADA, Barcelone, 28-29 mars.

GREMI (1990): Nouvelles formes d'organisation industrielle: réseaux d'innovation et milieux locaux. Actes du colloque GREMI, Neuchâtel, 11-12 novembre.

Håkansson, H. (1987): Industrial Technological Development. Croom Helm, London.

Håkansson, H. (1989): Corporate Technological Behaviour. Co-operation and Networks. Routledge, London, New York.

Håkansson, H./Johanson, J. (1993): The Network as a Governance Structure. Interfirm Cooperation Beyond Markets and Hierarchies. In: GRABHER, G. (Ed.) (1993): The Embedded Firm. On the Socioeconomics of Industrial Networks. Routledge, London, New York.

Hall, E. (1959): The Silent Language. Doubleday, Garden City, New York.

Hall, E. (1969): The Hidden Dimension. Doubleday, Garden City, New York.

Hardt, P. (1996): Organisation dienstleistungsorientierter Unternehmen. Deutscher Universitäts-Verlag (Gabler), Wiesbaden.

Héraud, J.-A. (1987): Interaction, information et apprentissage: l'auto-organisation en économie. Working Paper BETA, 8709, Strasbourg.

Héraud, J.-A. (1988): Routine et innovation dans un système économique auto-organisé. In: Fundamenta Scientae. Vol. 9, 2/3. pp. 375-391.

Héraud, J.-A. et al. (1993): Les réseaux économiques et technologiques de l'industrie alsacienne. BETA, Strasbourg.

Héraud, J.-A./Muller, E (1998): The Impact of Universities and Research Institutions Labs on the Creation and Diffusion of Innovation-Relevant Knowledge: the Case of the Upper-Rhine Valley. Paper presented at the 38th Congress of the European Regional Science Association, August 28-31 1998, Vienna.

Herden, R. (1992): Technologieorientierte Außenbeziehungen im betrieblichen Innovationsmanagement. Ergebnisse einer empirischen Untersuchung. Physica-Verlag, Heidelberg.

Hipp, C. (1998): The Role of Knowledge-intensive Business Services in the New Mode of Knowledge Production. Paper presented at the International Conference on Science, Technology and Society, 16-22 March 1998, Tokyo.

Homburg, C./Garbe, B. (1996): Industrielle Dienstleistungen. Bestandsaufnahme und Entwicklungsrichtungen. In: Zeitschrift für Betriebswirtschaft. 3. pp. 253-282.

Illeris, S. (1996): The Service Economy: A Geographical Approach. John Wiley & Sons Ltd, Baffins Lane.

Isard, W. (1956): Location and Space-Economy – A General Theory Relating to Industrial Location, Market Areas, Land Use, Trade, and Urban Structure. The MIT Press, Cambridge, MA.

Jaffe, A. (1989): Real Effects of Academic Research. In: American Economic Review. 79. pp. 957-970.

Jaffe, A./Trajtenberg, M./Henderson, R. (1993): Geographic Localization of Knowledge Spillovers as Evidenced by Patent Citations. In: Quarterly Journal of Economics. August. pp. 577-598.

Julien, P.-A. (1996): Information control: a key factor in small business development. Paper presented at the 41st ICSB world conference, June 17-19, Stockholm.

Kleinknecht, A. (1989): Firm Size and Innovation. Observations in Dutch Manufacturing Industries. SEO Reprint. 53. Amsterdam.

Kleinknecht, A./Poot, T.P. (1992): Do Regions Matter for R&D? In: Regional Studies. 26/3. pp. 221-232.

Kline, S./Rosenberg, N. (1986): An Overview of Innovation. In: Landau, R./Rosenberg, N. (Eds.) (1986): The Positive Sum Strategy: Harnessing Technology for Economic Growth. National Academy Press, Washington D.C. pp. 275-305.

Kœnig, G. (1993): Les Théories de la Firme. Economica, Paris.

König, H./Kukuk, M./Licht, G. (1996): Kooperationsverhalten von Unternehmen des Dienstleistungssektors. Internal report ZEW, Mannheim.

Koschatzky, K./Héraud, J.-A. (1996): Institutions of Technological Infrastructure. Final report to Eurostat on the project "Feasibility study on the statistical measurement of the Institutions of Technological Infrastructure". FhG-ISI, Karlsruhe and BETA (Université Louis Pasteur), Strasbourg.

Koschatzky, K./Muller, E. (1997): Firm Innovation and Region - Theoretical and Political Conclusions on Regional Innovation Networking. Paper presented at the 37th Congress of the European Regional Science Association, August 26-29 1997, Rome.

Krugman, P. (1991): Geography and Trade. The MIT Press, Cambridge.

Laborit, H. (1974): La nouvelle grille. Laffont, Paris.

Landau, R./Rosenberg, N. (Eds.) (1986): The Positive Sum Strategy: Harnessing Technology for Economic Growth. National Academy Press, Washington D.C.

Larue de Tournemine, R. (1991): Stratégies technologiques et processus d'innovation. Les Editions d'Organisation, Paris.

Le Bas, C./Torre, A. (1993): Survey sur les surveys d'innovation – Une première évaluation des enquêtes d'innovation européennes. In: Revue d'Economie Industrielle. 65. pp. 80-95.

Le Moigne, J.L. (1986): Vers un système d'information organisationnel. In: Revue Française de Gestion. Novembre -décembre. pp. 20-31.

Lichnerowicz, A./Perroux, F./Gadoffre, G. (Eds.): Information et communication. Maloine.

Lorenzen, M. (1998): Information Cost, Learning, and Trust - Lessons from Co-operation and Higher-order Capabilities Amongst Geographically Proximate Firms. DRUID Working Paper. No. 98-21. Copenhagen.

Lundvall, B.A. (1988): Innovation as an Interactive Process: From User-Producer Interaction to the National System of Innovation. In: Dosi, G./Freeman, C./Nelson, R./Silverberg, G./Soete, L. (Eds.) (1988): Technical Change and Economic Theory. Pinter Publishers, London. pp. 349-369.

Lundvall, B.A. (1992) (Ed.): National System of Innovation. Towards a Theory of Innovation and Interactive Learning. Pinter Publishers, London.

Mackun, P./Macpherson, A. (1997): Externally-assisted Product Innovation in the Manufacturing Sector: The Role of Location, In-house R&D and Outside Technical Support. In: Regional Studies. Vol. 31.7. pp. 659-668.

Macleod, R. (Ed.) (1986): Technology and the Human Prospect. Pinter Publishers, London.

Macpherson, A. (1997): The Role of Producer Service Outsourcing in the Innovation Performance of New York State Manufacturing Firms. In: Annals of the Association of American Geographers. 87. pp. 52-71.

Magidson, J./SPSS Inc. (1993): SPSS for Windows - CHAID Release 6.0. SPSS, Chicago.

Maillat, D. (Ed.) (1992): Nouvelles formes d'organisation industrielle: réseaux d'innovation et milieux locaux. EDES, Neuchâtel.

Maillat, D./Perrin, J.C. (Eds.) (1992): Entreprises innovatrices et développement territorial. EDES, Neuchâtel.

Malecki, E.J. (1990): Promoting and Inhibiting Factors in the Regional Environmental System. In: Ewers, H.J./Allesch, J. (Eds.) (1990): Innovation and Regional Development. Strategies, Instruments and Policy Coordination. Walter de Gruyter, Berlin, New York. pp. 123-147.

Männel, W. (1981): Die Wahl zwischen Eigenfertigung und Fremdbezug. C.E. Poeschel Verlag, Stuttgart.

Marshall, A. (1900): Elements of Economics of Industry. Macmillan, London.

Marshall, J.N. (1982): Linkages between manufacturing industry and business services. In: Environment and Planning. 14. pp. 1523-1540.

Martinelli, F./Schoenberger, E. (1992): Les oligopoles se portent bien, merci! In: Benko, G./Lipietz, A. (1992): Les régions qui gagnent - Districts et réseaux: les nouveaux paradigmes de la géographie économique. Presses Universitaires de France, Paris. pp. 163-188.

Massard, N. (1991): L'industrialisation des nouvelles technologies: le cas des fibres optiques. Presses Universitaires de Lyon, Lyon.

Mckelvey, M. (1998): Evolutionary innovations: learning, entrepreneurship and the dynamics of the firm. In: Journal of Evolutionary Economics. 8. pp. 157-175.

Miles, I./Kastrinos, N./Flanagan, K./Bildebeek, R./Den Hertog, P./Huntink, W./Bouman, M. (1994): Knowledge Intensive Business Services: Their Roles as Users, Carriers and Sources of Innovation. PREST, Manchester.

Miller, R. (1996): Towards the Knowledge Economy: New Institutions for Human Capital Accounting. In: OECD (1996): Employment and Growth in the Knowledge-based Economy. OECD Documents, Paris. pp. 69-80.

Morgan, K. (1997): The Learning Region: Institutions, Innovation and Regional Renewal. In: Regional Studies. 31.5. pp. 491-503.

Morin, J. (1986): Le management des ressources technologiques: un vecteur de l'innovation. In: Revue Française de Gestion. Sept./Oct. pp. 31-38.

Muller, E./Zenker, A./Meyer-Krahmer, F. (1998): The Consequences of a Growing Codification of Knowledge on the Evolution Capacities of European Firms and Regions. Working document FhG-ISI, Karlsruhe.

Mustar, P./Callon, M. (1992): Réseaux de l'innovation. In: Curien, N. (Ed.) (1992): Économie et Management des entreprises de réseau. Economica, Paris. pp. 115-128.

Nelson, R./Winter, S. (1974): Neoclassical vs. Evolutionary Theories of Economic Growth. Critique and Prospectus. In: Economic Journal. December, pp. 886-905.

Nelson, R./Winter, S. (1975): Growth Theory from an Evolutionary Perspective: The Differential Productivity Puzzle. In: The American Economic Review. 65/2. pp. 338-344.

Nelson, R./Winter, S. (1977): In Search of a Useful Theory of Innovation. In: Research Policy. 6. pp. 36-76.

Nielsen, N.C. (1996): The Concept of Technological Service Infrastructures: Innovation and the Creation of Good Jobs. In: OECD (1996): Employment and Growth in the Knowledge-based Economy. OECD Documents, Paris. pp. 237-254.

Nonaka, I. (1994): A Dynamic Theory of Organizational Knowledge Creation. In: Organization Science. 5/1. pp. 14-37.

OCDE (1993): Les Petites et Moyennes Entreprises: technologie et compétitivité. OCDE, Paris.

OECD (1994): Proposed Standard Practice for Surveys of Research and Experimental Development - Frascati Manual, fifth edition. OECD, Paris.

OECD (1996): Employment and Growth in the Knowledge-based Economy. OECD Documents, Paris.

OECD (1997): Proposed Guidelines for Collecting and Interpreting Technological Innovation Data - Oslo Manual, second edition. OECD, Paris.

Ohmae, K. (1993): The Rise of the Region State. In: Foreign Affairs. 72. pp. 78-87.

Pavitt, K. (1984): Sectoral Patterns of Technical Change: Towards a Taxonomy and a Theory. In: Research Policy. 13. pp. 343-373.

Pecqueur, B. (1989): Le développement local. Syros, Paris.

Perrez, C. (1983): Structural Change and Assimilation of New Technologies in the Economic and Social System. In: Futures. 15/5. pp. 357-375.

Perrin, J.-C. (1990): Organisation industrielle: La composante territoriale. In: Revue d'Economie Industrielle. 51. pp. 276-303.

Phillips, A. (1971): Technology and Market Structure. Lexington Books, Lexington.

Pilorget, L. (1994): France - Allemagne. Comment promouvoir la coopération industrielle entre PME? Peter Lang Verlag, Frankfurt.

Piore, M./Sabel, C. (1984): The Second Industrial Divide: Possibilities for Prosperity. Basic Books, New-York,.

Polanyi, M. (1966): The Tacit Dimension. Doubleday, New York.

Porter, M. (1990): The Competitive Advantage of Nations. Macmillan, London.

Rallet, A. (1993): Choix de proximité et processus d'innovation technologique. In: Revue d'Economie Régionale et Urbaine. 3. pp. 365-385.

Reich, R. (1991): The Work of Nations. Preparing Ourselves for 21st-Century Capitalism. Alfred A. Knopf, New York.

Rifkin, J. (1996): La fin du travail. Éditions La Découverte, Paris.

Rosenberg, N. (1976): Perspective on Technology. Harvard University Press.

Rosenberg, N. (1982): Inside the Black Box: Technology and Economics. University Press, Cambridge.

Rosenberg, N. (1990): Why do firms do basic research (with their own money)? In: Research Policy. 19. pp. 165-174.

Saviotti, P.P. (1998):On the dynamics of appropriability, of tacit and of codified knowledge. In: Research Policy. 26. pp. 843-856

Schmoch, U./Hinze, S./Jäckel, G./Kirsch, N./Meyer-Krahmer, F./Münt, G. (1993): Constraints and Opportunities for the Dissemination and Exploitation of R&D Activities: The R&D Environment - Structural Features and their Modelling. FhG-ISI, Karlsruhe.

Schumpeter, J.A. (1935): Theorie der wirtschaftlichen Entwicklung - Eine Untersuchung über Unternehmergewinn, Kapital, Kredit, Zins und den Konjunkturzyklus. Duncker & Humblot, München, Leipzig.

Schumpeter, J.A. (1950): Capitalism, Socialism and Democracy. Harper and Brothers, New-York.

Segal Quince Wicksteed Limited (1991): Evaluation of the Consultancy Initiatives - Third stage. HMSO, London.

Shostack, L.G. (1977): Breaking Free from Product Marketing. In: Journal of Marketing. 41. pp. 73-80.

Simon, H. (1982): Models for Bounded Rationality: Behavioural Economics and Business Organization. The MIT Press, Cambridge, London.

Soete, L./Miozzo, M. (1990): Trade and Development in Services: a Technological Perspective. MERIT, Maastricht.

Staudacher, C. (1991): Dienstleistungen; Raumstruktur und räumliche Prozesse – Eine Einführung in die Dienstleistungsgeographie. Service-Fachverlag, Wien.

Storper, M. (1997): The Regional World. Territorial Development in a Global Economy. The Guilford Press, New York.

Strambach, S. (1997): Wissensintensive unternehmensorientierte Dienstleistungen - ihre Bedeutung für die Innovations- und Wettbewerbsfähigkeit Deutschlands. In: Deutsches Institut für Wirtschaftsforschung (Ed.). Vierteljahreshefte zur Wirtschaftsforschung. 2. pp. 230-242.

Strambach, S. (1998): Knowledge-Intensive Business Services (KIBS) as an Element of Learning Regions - the Case of Baden-Württemberg. Paper presented at the ERSA Conference, August 28-31 1998, Vienna.

Teece, D.J. (1986): Profiting from Technological Innovation: Implications for Integration, Collaboration, Licensing and Public Policy. In: Research Policy. 15. pp. 285-305.

Tomlinson, M. (1997): The Contribution of Services to Manufacturing Industry: Beyond the Deindustrialisation Debate. CRIC Discussion Paper No 5. CRIC, Manchester.

Tunney, J. (1998): The Need of Strategic Awareness of European Union Law for HTSFs. Paper presented at the 6th Annual International Conference on High-Tech Small Firms. University of Twente (The Netherlands). 4 and 5 June 1998.

Veltz, P. (1992): Hiérarchies et réseaux dans l'organisation de la production et du territoire. In: BENKO, G./LIPIETZ, A. (Eds.) (1992): Les régions qui gagnent - Districts et réseaux: les nouveaux paradigmes de la géographie économique. Presses Universitaires de France, Paris. pp. 293-313.

Von Einem, E./Helmstädter H.G. (1994): Produktinnovationen in Wechselbeziehungen zwischen Industrie und Dienstleistungen. Thesenpapier zum Kolloquium im Rahmen des DFG Schwerpunktprogramms "Technologischer Wandel und Regionalentwicklung in Europa". IfS, Bonn.

Von Hippel, E. (1988): The Sources of Innovation. Oxford University Press, New York.

Voss, C.A./Johnson, R./Silvestro, R./Fitzgerald, L./Brignall, T.J. (1992): Measurement of Innovation and Design Performance in Services. In: Design Management Journal. Winter 1992. pp. 40-46.

Weber, A. (1929): Theory of the Location of Industry. University of Chicago Press, Chicago.

Williamson, O.E. (1975): Markets and Hierarchies: Analysis and Antitrust Implications. The Free Press, New York.

Williamson, O.E. (1981): The Economics of Organization: The Transaction Cost Approach. In: American Journal of Sociology. 87/3. pp. 548-577.

Wood, P. (1998): The Rise of Consultancy and the Prospect for Regions. Paper presented at the 38th Congress of the European Regional Science Association, August 28-31 1998, Vienna.

ZEW/FhG ISI (1999): Services in the Future – Innovation Activities in the Service Sector (Executive Summary, Survey 1997). Infas, Mannheim.

Appendix

Appendix A: Basic frequencies of the selected variables

Variable: SECTOR (SMEs Sample)

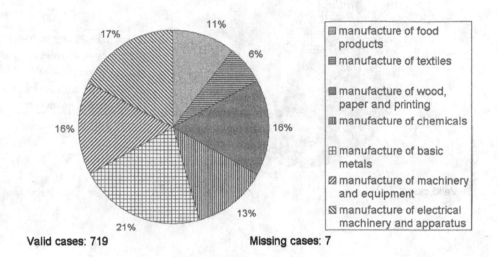

Valid cases: 719   Missing cases: 7

Variable: SECTOR (KIBS Sample)

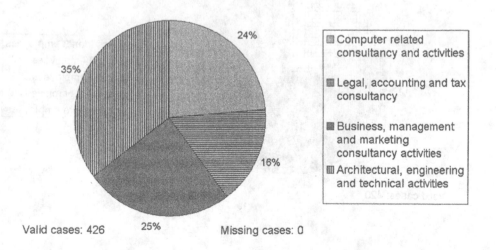

Valid cases: 426   25%   Missing cases: 0

Variable: SIZE (SMEs Sample)

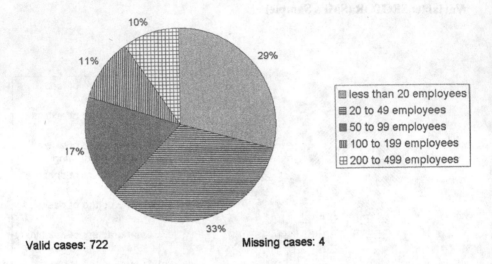

Valid cases: 722   Missing cases: 4

Variable: SIZE (KIBS Sample)

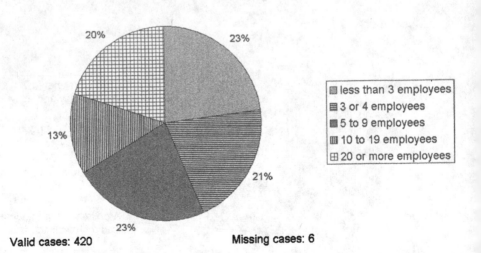

Valid cases: 420   Missing cases: 6

Variable: COUNT (SMEs Sample)

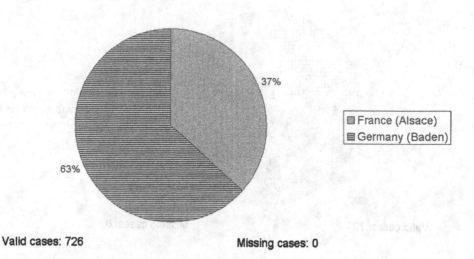

37%

63%

France (Alsace)
Germany (Baden)

Valid cases: 726   Missing cases: 0

Variable: COUNT (KIBS Sample)

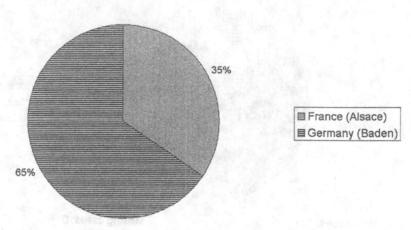

35%

65%

France (Alsace)
Germany (Baden)

Valid cases: 426   Missing cases: 0

Variable: REGTYP (SMEs Sample)

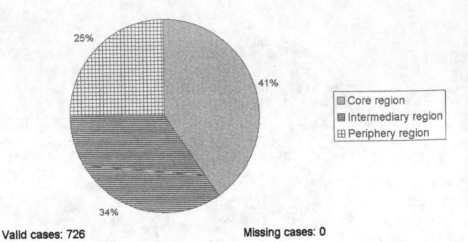

Valid cases: 726   Missing cases: 0

Variable: REGTYP (KIBS Sample)

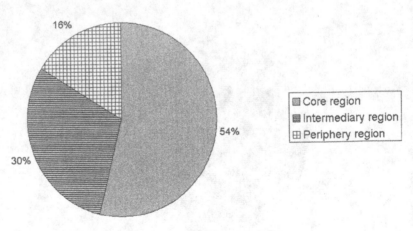

Valid cases: 426   Missing cases: 0

Variable: IKIBS (SMEs Sample)

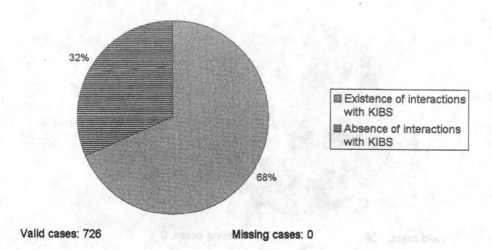

Valid cases: 726   Missing cases: 0

Variable: ISMES (KIBS Sample)

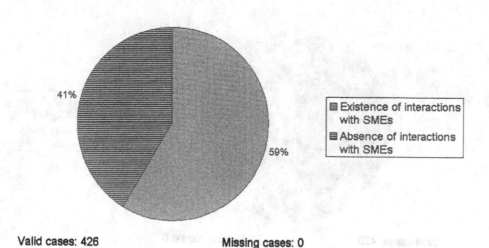

Valid cases: 426   Missing cases: 0

Variable: PKIBS (SMEs Sample)

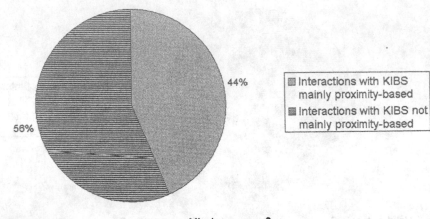

Valid cases: 726   Missing cases: 0

Variable: PSMES (KIBS Sample)

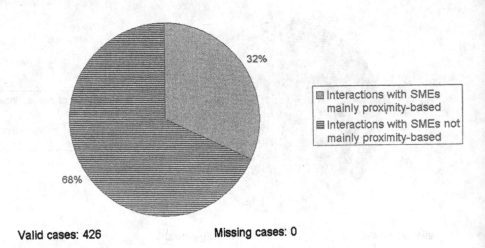

Valid cases: 426   Missing cases: 0

Variable: NITI (SMEs Sample)

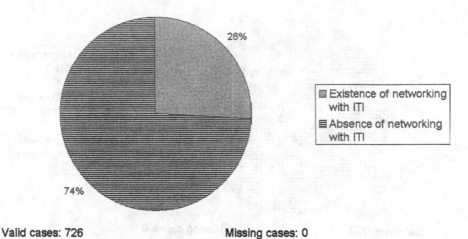

Valid cases: 726   Missing cases: 0

Variable: NITI (KIBS Sample)

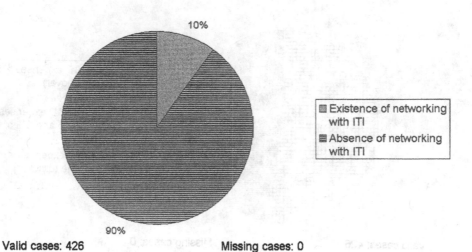

Valid cases: 426   Missing cases: 0

Variable: LEVRD (SMEs Sample)

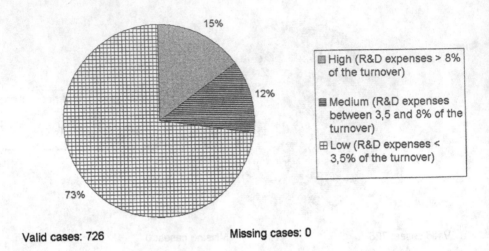

Valid cases: 726   Missing cases: 0

Variable: LEVRD (KIBS Sample)

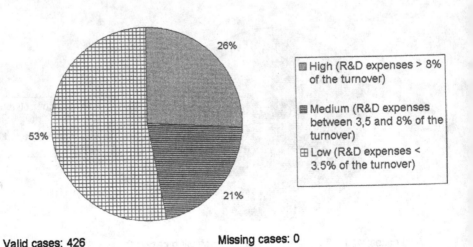

Valid cases: 426   Missing cases: 0

Variable: GROWTH (SMEs Sample)

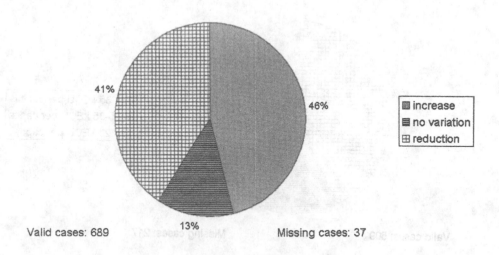

41%

46%

13%

- increase
- no variation
- reduction

Valid cases: 689   Missing cases: 37

Variable: GROWTH (KIBS Sample)

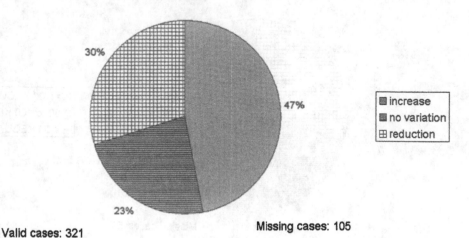

30%

47%

23%

- increase
- no variation
- reduction

Valid cases: 321   Missing cases: 105

Variable: NETSALES (SMEs Sample)

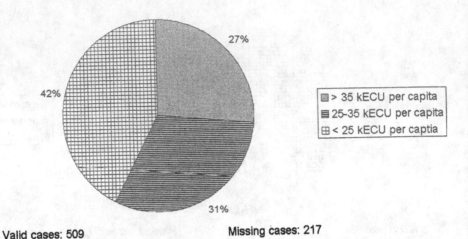

Valid cases: 509   Missing cases: 217

Variable: TURNOVER (KIBS Sample)

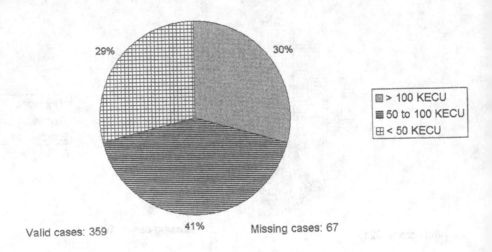

Valid cases: 359   Missing cases: 67

Variable: LEVEXP (SMEs Sample)

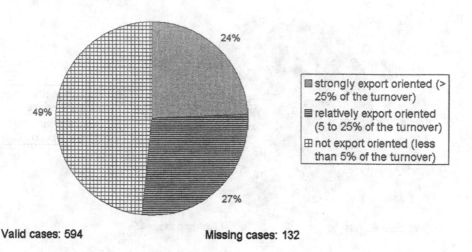

Valid cases: 594   Missing cases: 132

Variable: LEVEXP (KIBS Sample)

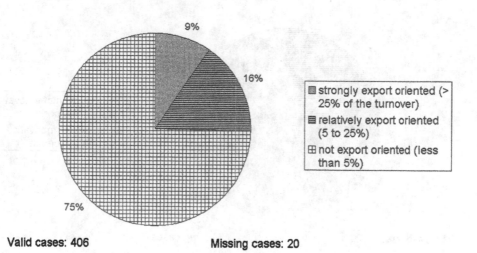

Valid cases: 406   Missing cases: 20

Variable: INNOV (SMEs Sample)

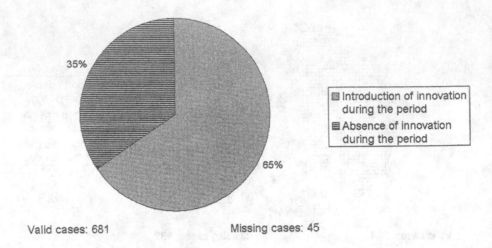

Valid cases: 681   Missing cases: 45

Variable: INNOV (KIBS Sample)

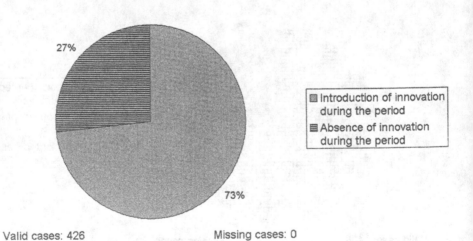

Valid cases: 426   Missing cases: 0

Appendix B: Selected bivariate tests

SMEs sample χ^2-Test (Pearson) GROWTH by:	Value	Degrees of freedom	Significance [*:at 5% level, **: at 1% level]
INNOV	12,77157	2	**
LEVRD	10,15618	4	*
IKIBS	0,23235	2	-
PKIBS	1,1,5973	2	-
NITI	4,43887	2	-
SIZE	35,81691	8	**
SECTOR	41,05741	12	**
COUNT	4,59520	2	-
REGTYP	1,98359	4	-

KIBS sample χ^2-Test (Pearson) GROWTH by:	Value	Degrees of freedom	Significance [*:at 5% level, **: at 1% level]
INNOV	37,46486	2	**
LEVRD	12,99702	4	*
ISMES	10,70936	2	**
PSMES	3,19359	2	-
NITI	1,85266	2	-
SIZE	72,79453	8	**
SECTOR	9,26184	6	-
COUNT	1,37015	2	-
REGTYP	7,75717	4	-

SMEs sample χ^2-Test (Pearson) NETSALES by:	Value	Degrees of freedom	Significance [*:at 5% level, **: at 1% level]
INNOV	8,78999	2	*
LEVRD	3,16262	4	-
IKIBS	11,50073	2	**
PKIBS	0,52715	2	-
NITI	3,36592	2	-
SIZE	6,85304	8	-
SECTOR	29,81120	12	**
COUNT	5,23199	2	-
REGTYP	8,17078	4	-

KIBS sample χ^2-Test (Pearson) TURNOVER by:	Value	Degrees of freedom	Significance [*:at 5% level, **: at 1% level]
INNOV	1,33998	2	-
LEVRD	5,17974	4	-
ISMES	2,75748	2	-
PSMES	0,76561	2	-
NITI	0,20646	2	-
SIZE	4,88396	8	-
SECTOR	31,26897	6	**
COUNT	2,37332	2	-
REGTYP	9,04283	4	-

SMEs sample χ²-Test (Pearson) LEVEXP by:	Value	Degrees of freedom	Significance [*:at 5% level, **: at 1% level]
INNOV	21,12714	2	**
LEVRD	28,95715	4	**
IKIBS	6,25704	2	*
PKIBS	4,35165	2	-
NITI	20,85182	2	**
SIZE	58,35306	8	**
SECTOR	69,04362	12	**
COUNT	9,07261	2	*
REGTYP	3,03496	4	-

KIBS sample χ²-Test (Pearson) LEVEXP by:	Value	Degrees of freedom	Significance [*:at 5% level, **: at 1% level]
INNOV	11,19349	2	**
LEVRD	18,72102	4	**
ISMES	14,03539	2	**
PSMES	11,42542	2	**
NITI	7,98185	2	*
SIZE	30,95680	8	**
SECTOR	12,99419	6	*
COUNT	2,53597	2	-
REGTYP	3,60633	4	-

SMEs sample χ^2-Test (Pearson) INNOV by:	Value	Degrees of freedom	Significance [*:at 5% level, **: at 1% level]
LEVRD	146,04980	2	**
IKIBS	33,55175	1	**
PKIBS	9,79592	1	**
NITI	48,42509	1	**
SIZE	69,98648	4	**
SECTOR	38,52068	6	**
COUNT	6,67216	1	**
REGTYP	1,75380	2	-

KIBS sample χ^2-Test (Pearson) INNOV by:	Value	Degrees of freedom	Significance [*:at 5% level, **: at 1% level]
LEVRD	137,74677	2	**
ISMES	33,14083	1	**
PSMES	14,81227	1	**
NITI	17,02464	1	**
SIZE	63,88945	4	**
SECTOR	8,51786	3	*
COUNT	3,11121	1	-
REGTYP	0,84187	2	-

SMEs sample χ²-Test (Pearson) LEVRD by:	Value	Degrees of freedom	Significance [*:at 5% level, **: at 1% level]
IKIBS	48,36979	2	**
PKIBS	12,81158	2	**
NITI	43,47736	2	**
SIZE	43,39502	8	**
SECTOR	107,81446	12	**
COUNT	27,55506	2	**
REGTYP	18,60024	4	**

KIBS sample χ²-Test (Pearson) LEVRD by:	Value	Degrees of freedom	Significance [*:at 5% level, **: at 1% level]
ISMES	44,28713	2	**
PSMES	6,53596	2	*
NITI	12,63159	2	**
SIZE	32,43115	8	**
SECTOR	30,02400	6	**
COUNT	20,62338	2	**
REGTYP	5,58377	4	-

SMEs sample χ^2-Test (Pearson) NITI by:	Value	Degrees of freedom	Significance [*:at 5% level, **: at 1% level]
IKIBS	32,94454	1	**
PKIBS	0,07558	1	-
SIZE	113,51450	4	**
SECTOR	21,27673	6	*
COUNT	0,09735	1	-
REGTYP	0,56059	2	-

KIBS sample χ^2-Test (Pearson) NITI by:	Value	Degrees of freedom	Significance [*:at 5% level, **: at 1% level]
ISMES	5,88871	1	*
PSMES	0,81626	1	-
SIZE	11,14952	4	*
SECTOR	4,02434	3	-
COUNT	12,86834	1	**
REGTYP	1,43701	2	-

SMEs sample χ^2-Test (Pearson) IKIBS by:	Value	Degrees of freedom	Significance [*:at 5% level, **: at 1% level]
SIZE	22,70945	4	**
SECTOR	3,77816	6	-
COUNT	43,79002	1	**
REGTYP	10,00126	2	**

KIBS sample χ^2-Test (Pearson) ISMES by:	Value	Degrees of freedom	Significance [*:at 5% level, **: at 1% level]
SIZE	18,08945	4	**
SECTOR	35,85138	3	**
COUNT	7,54119	1	**
REGTYP	2,13719	2	-

SMEs sample χ²-Test (Pearson) PKIBS by:	Value	Degrees of freedom	Significance [*:at 5% level, **: at 1% level]
SIZE	3,65447	4	-
SECTOR	3,00700	6	-
COUNT	15,32247	1	**
REGTYP	3,47743	2	-

KIBS sample χ²-Test (Pearson) PSMES by:	Value	Degrees of freedom	Significance [*:at 5% level, **: at 1% level]
SIZE	10,12909	4	*
SECTOR	8,58870	3	*
COUNT	0,73801	1	-
REGTYP	2,47179	2	-

Appendix C: PROBIT analysis

C1   The SMEs sample

probit growth levexp netsales innov levrd ikibs pkibs niti size count regtyp sector

```
Iteration 0:  Log Likelihood =-497.63196
Iteration 1:  Log Likelihood =-489.90571
Iteration 2:  Log Likelihood =-489.90295
```

Probit Estimates

```
                                                    Number of obs =     726
                                                    chi2(11)      =   15.46
                                                    Prob > chi2   = 0.1625
Log Likelihood = -489.90295                         Pseudo R2     = 0.0155
```

growth	Coef.	Std. Err.	z	P>|z|	[95% Conf. Interval]	
levexp	-.0264581	.1239865	-0.213	0.831	-.2694672	.216551
netsales	-.0278085	.1237584	-0.225	0.822	-.2703706	.2147536
innov	.187143	.1166413	1.604	0.109	-.0414698	.4157558
levrd	.1439793	.1276901	1.128	0.260	-.1062887	.3942473
ikibs	-.0197486	.139119	-0.142	0.887	-.2924169	.2529197
pkibs	.0928085	.1212172	0.766	0.444	-.144773	.3303899
niti	-.1367577	.1214217	-1.126	0.260	-.3747398	.1012245
size	-.2095225	.1298739	-1.613	0.107	-.4640706	.0450256
count	.2336407	.1136555	2.056	0.040	.01088	.4564015
regtyp	-.1864015	.1212427	-1.537	0.124	-.4240327	.0512298
sector	-.0971811	.1104876	-0.880	0.379	-.3137328	.1193705
_cons	-.1665875	.1345765	-1.238	0.216	-.4303526	.0971777

probit levexp growth netsales innov levrd ikibs pkibs niti size count regtyp sector

```
Iteration 0:  Log Likelihood = -361.6224
Iteration 1:  Log Likelihood =-329.15619
Iteration 2:  Log Likelihood =-328.88532
Iteration 3:  Log Likelihood =-328.88522
```

Probit Estimates

```
                                                    Number of obs =     726
                                                    chi2(11)      =   65.47
                                                    Prob > chi2   = 0.0000
Log Likelihood = -328.88522                         Pseudo R2     = 0.0905
```

levexp	Coef.	Std. Err.	z	P>|z|	[95% Conf. Interval]	
growth	-.0235864	.1134153	-0.208	0.835	-.2458763	.1987034
netsales	.4521262	.1354239	3.339	0.001	.1867003	.7175521
innov	.0734648	.1415536	0.519	0.604	-.203975	.3509047
levrd	.1206269	.1457066	0.828	0.408	-.1649528	.4062067
ikibs	.2596041	.1623882	1.599	0.110	-.058671	.5778792
pkibs	-.2512962	.1379451	-1.822	0.068	-.5216637	.0190712
niti	-.1301741	.141585	-0.919	0.358	-.4076756	.1473274
size	.2292541	.1450866	1.580	0.114	-.0551104	.5136185
count	.2525239	.1351763	1.868	0.062	-.0124168	.5174647
regtyp	-.0205957	.1445533	-0.142	0.887	-.303915	.2627236
sector	.7035514	.1217544	5.778	0.000	.4649172	.9421855
_cons	-1.46276	.1762393	-8.300	0.000	-1.808183	-1.117338

probit netsales levexp growth innov levrd ikibs pkibs niti size count regtyp sector

```
Iteration 0:   Log Likelihood =-348.69595
Iteration 1:   Log Likelihood =-332.92009
Iteration 2:   Log Likelihood =-332.81581
Iteration 3:   Log Likelihood =-332.81579
```

Probit Estimates

```
                                                Number of obs  =    726
                                                chi2(11)       =  31.76
                                                Prob > chi2    = 0.0008
Log Likelihood = -332.81579                     Pseudo R2      = 0.0455
```

netsales	Coef.	Std. Err.	z	P>|z|	[95% Conf. Interval]	
levexp	.4561824	.1351723	3.375	0.001	.1912495	.7211152
growth	-.0241915	.1125004	-0.215	0.830	-.2446882	.1963051
innov	.1802831	.136172	1.324	0.186	-.0866091	.4471754
levrd	-.1625968	.1463575	-1.111	0.267	-.4494522	.1242586
ikibs	.514356	.1627816	3.160	0.002	.1953099	.8334021
pkibs	-.1658073	.1340069	-1.237	0.216	-.428456	.0968414
niti	.1276284	.1377912	0.926	0.354	-.1424374	.3976942
size	-.2486895	.15403	-1.615	0.106	-.5505827	.0532038
count	-.053723	.1324467	-0.406	0.685	-.3133138	.2058677
regtyp	.119716	.1421595	0.842	0.400	-.1589114	.3983435
sector	-.1444223	.1314677	-1.099	0.272	-.4020943	.1132498
_cons	-1.364866	.1738245	-7.852	0.000	-1.705556	-1.024176

probit innov levrd ikibs pkibs niti size count regtyp sector

```
Note: levrd~=0 predicts success perfectly
      levrd dropped and 197 obs not used

Iteration 0:   Log Likelihood =-365.76602
Iteration 1:   Log Likelihood =-326.61253
Iteration 2:   Log Likelihood = -326.1773
Iteration 3:   Log Likelihood =-326.17704
```

Probit Estimates

```
                                                Number of obs  =    529
                                                chi2(7)        =  79.18
                                                Prob > chi2    = 0.0000
Log Likelihood = -326.17704                     Pseudo R2      = 0.1082
```

innov	Coef.	Std. Err.	z	P>|z|	[95% Conf. Interval]	
ikibs	.4994687	.1612672	3.097	0.002	.1833907	.8155466
pkibs	.0932864	.1532947	0.609	0.543	-.2071657	.3937386
niti	.6324277	.158229	3.997	0.000	.3223045	.9425509
size	.6622936	.1742468	3.801	0.000	.3207762	1.003811
count	.1072515	.1340611	0.800	0.424	-.1555035	.3700064
regtyp	-.0117776	.1550419	-0.076	0.939	-.3156541	.2920989
sector	.2810334	.136951	2.052	0.040	.0126144	.5494524
_cons	-.734153	.1559307	-4.708	0.000	-1.039772	-.4285345

probit levrd ikibs pkibs niti size count regtyp sector

```
Iteration 0:  Log Likelihood = -424.4173
Iteration 1:  Log Likelihood = -348.2112
Iteration 2:  Log Likelihood = -345.6933
Iteration 3:  Log Likelihood =-345.67479
Iteration 4:  Log Likelihood =-345.67479
```

Probit Estimates

```
                                          Number of obs =     726
                                          chi2(7)       = 157.49
                                          Prob > chi2   = 0.0000
Log Likelihood = -345.67479               Pseudo R2     = 0.1855
```

---

levrd	Coef.	Std. Err.	z	P>|z|	[95% Conf. Interval]	
ikibs	.7375006	.1642424	4.490	0.000	.4155913	1.05941
pkibs	.0232757	.1300986	0.179	0.858	-.2317129	.2782643
niti	.413342	.1282257	3.224	0.001	.1620243	.6646597
size	.3324186	.1383681	2.402	0.016	.061222	.6036151
count	-.288782	.1343281	-2.150	0.032	-.5520602	-.0255038
regtyp	-.1483937	.1327111	-1.118	0.263	-.4085027	.1117152
sector	.8107999	.1136462	7.134	0.000	.5880574	1.033542
_cons	-1.477929	.1649427	-8.960	0.000	-1.80121	-1.154647

---

probit ikibs niti size count regtyp sector

```
Iteration 0:  Log Likelihood =-454.10673
Iteration 1:  Log Likelihood =-412.13479
Iteration 2:  Log Likelihood =-411.55381
Iteration 3:  Log Likelihood =-411.55309
```

Probit Estimates

```
                                          Number of obs =     726
                                          chi2(5)       =  85.11
                                          Prob > chi2   = 0.0000
Log Likelihood = -411.55309               Pseudo R2     = 0.0937
```

---

ikibs	Coef.	Std. Err.	z	P>|z|	[95% Conf. Interval]	
niti	.6707749	.1335722	5.022	0.000	.4089781	.9325717
size	.3125476	.1438969	2.172	0.030	.0305148	.5945804
count	-.6971617	.1173189	-5.942	0.000	-.9271025	-.4672209
regtyp	.0077425	.1345739	0.058	0.954	-.2560174	.2715024
sector	-.1338844	.1116785	-1.199	0.231	-.3527702	.0850014
_cons	.5853824	.1139643	5.137	0.000	.3620164	.8087483

---

probit pkibs niti size count regtyp sector

```
Iteration 0:   Log Likelihood =-497.12234
Iteration 1:   Log Likelihood =-488.99401
Iteration 2:   Log Likelihood =-488.99191
Iteration 3:   Log Likelihood =-488.99191
```

Probit Estimates   Number of obs =   726
                                                      chi2(5)   = 16.26
                                                      Prob > chi2   = 0.0061
Log Likelihood = -488.99191   Pseudo R2   = 0.0164

pkibs	Coef.	Std. Err.	z	P>|z|	[95% Conf. Interval]	
niti	.0621194	.1153937	0.538	0.590	-.1640481	.288287
size	-.1016474	.1264205	-0.804	0.421	-.349427	.1461323
count	-.3800612	.110407	-3.442	0.001	-.5964548	-.1636675
regtyp	-.015481	.1196438	-0.129	0.897	-.2499785	.2190165
sector	-.0035422	.1030774	-0.034	0.973	-.2055701	.1984858
_cons	-.0095921	.1038094	-0.092	0.926	-.2130548	.1938707

probit niti ikibs pkibs count size sector regtyp

```
Iteration 0:   Log Likelihood =-414.18731
Iteration 1:   Log Likelihood =-352.44452
Iteration 2:   Log Likelihood =-351.74283
Iteration 3:   Log Likelihood = -351.7419
```

Probit Estimates   Number of obs =   726
                                                      chi2(6)   = 124.89
                                                      Prob > chi2   = 0.0000
Log Likelihood = -351.7419   Pseudo R2   = 0.1508

niti	Coef.	Std. Err.	z	P>|z|	[95% Conf. Interval]	
ikibs	.868817	.1517942	5.724	0.000	.571306	1.166328
pkibs	-.3592631	.1274199	-2.820	0.005	-.6090015	-.1095247
count	.1816331	.1288793	1.409	0.159	-.0709656	.4342318
size	.9459261	.1256349	7.529	0.000	.6996861	1.192166
sector	.3435802	.1146756	2.996	0.003	.1188203	.5683402
regtyp	-.0555008	.1382873	-0.401	0.688	-.3265389	.2155374
_cons	-1.503876	.1621525	-9.274	0.000	-1.821689	-1.186063

C2   The KIBS sample

`probit growth levexp turnover innov levrd ismes psmes niti size count regtyp sector`

```
Iteration 0:   Log Likelihood =-276.96985
Iteration 1:   Log Likelihood =-235.97357
Iteration 2:   Log Likelihood =-235.05285
Iteration 3:   Log Likelihood =-235.04971
Iteration 4:   Log Likelihood =-235.04971
```

Probit Estimates

```
                                              Number of obs =     426
                                              chi2(11)      =   83.84
                                              Prob > chi2   = 0.0000
Log Likelihood = -235.04971                   Pseudo R2     = 0.1514
```

growth	Coef.	Std. Err.	z	P>|z|	[95% Conf. Interval]	
levexp	.2586918	.1634324	1.583	0.113	-.0616298	.5790134
turnover	.2935072	.1563021	1.878	0.060	-.0128393	.5998537
innov	.5730635	.2051875	2.793	0.005	.1709033	.9752237
levrd	-.0407501	.163849	-0.249	0.804	-.3618882	.2803881
ismes	-.1726563	.1867892	-0.924	0.355	-.5387564	.1934438
psmes	.2589506	.1824663	1.419	0.156	-.0986768	.616578
niti	.0759467	.2198264	0.345	0.730	-.3549052	.5067985
size	.9147817	.1472782	6.211	0.000	.6261218	1.203442
count	.2614375	.1553814	1.683	0.092	-.0431045	.5659796
regtyp	.2225584	.2054467	1.083	0.279	-.1801098	.6252266
sector	-.0933378	.1409467	-0.662	0.508	-.3695883	.1829127
_cons	-1.688821	.2547009	-6.631	0.000	-2.188026	-1.189617

`probit levexp growth turnover innov levrd ismes psmes niti size count regtyp sector`

```
Iteration 0:   Log Likelihood =-235.63238
Iteration 1:   Log Likelihood =-200.83601
Iteration 2:   Log Likelihood = -200.2099
Iteration 3:   Log Likelihood =-200.20842
Iteration 4:   Log Likelihood =-200.20842
```

Probit Estimates

```
                                              Number of obs =     426
                                              chi2(11)      =   70.85
                                              Prob > chi2   = 0.0000
Log Likelihood = -200.20842                   Pseudo R2     = 0.1503
```

levexp	Coef.	Std. Err.	z	P>|z|	[95% Conf. Interval]	
growth	.2537634	.1576095	1.610	0.107	-.0551456	.5626723
turnover	.2841767	.1630888	1.742	0.081	-.0354715	.6038248
innov	.0027216	.2273639	0.012	0.990	-.4429035	.4483466
levrd	.3752691	.1818668	2.063	0.039	.0188168	.7317214
ismes	.7830308	.1822757	4.296	0.000	.425777	1.140285
psmes	-.7088483	.1817114	-3.901	0.000	-1.064996	-.3527006
niti	.2211765	.2301666	0.961	0.337	-.2299418	.6722947
size	.3934704	.1634981	2.407	0.016	.0730201	.7139207
count	-.0389084	.1672547	-0.233	0.816	-.3667215	.2889048
regtyp	.3562133	.2201699	1.618	0.106	-.0753118	.7877384
sector	-.2999891	.1498363	-2.002	0.045	-.5936628	-.0063154
_cons	-1.749811	.2703447	-6.473	0.000	-2.279677	-1.219946

```
probit turnover levexp growth innov levrd ismes psmes niti size count regtyp sector

Iteration 0:  Log Likelihood = -239.0039
Iteration 1:  Log Likelihood =-226.98888
Iteration 2:  Log Likelihood =-226.93761
Iteration 3:  Log Likelihood =-226.93761

Probit Estimates                                     Number of obs =     426
                                                     chi2(11)      =   24.13
                                                     Prob > chi2   = 0.0122
Log Likelihood = -226.93761                          Pseudo R2     = 0.0505

------------------------------------------------------------------------------
turnover |    Coef.    Std. Err.      z     P>|z|      [95% Conf. Interval]
---------+--------------------------------------------------------------------
  levexp |  .2966816   .1657048     1.790   0.073    -.0280939    .6214571
  growth |  .2641894   .1537591     1.718   0.086     -.037173    .5655518
   innov |  .1927117   .2052283     0.939   0.348    -.2095284    .5949518
   levrd | -.1020388   .1718836    -0.594   0.553    -.4389245    .2348469
   ismes |  .2285387   .1813061     1.261   0.207    -.1268147    .5838921
   psmes | -.2574605   .1813019    -1.420   0.156    -.6128057    .0978847
    niti | -.1236488   .2328159    -0.531   0.595    -.5799596     .332662
    size | -.2073791   .1564588    -1.325   0.185    -.5140326    .0992745
   count | -.2287497   .1580951    -1.447   0.148    -.5386105     .081111
  regtyp |  .1059018   .1975012     0.536   0.592    -.2811934     .492997
  sector |  .3970765   .1406232     2.824   0.005     .1214601     .672693
   _cons | -1.107467    .231383    -4.786   0.000    -1.560969   -.6539642
------------------------------------------------------------------------------

probit innov levrd ismes psmes niti size count regtyp sector

Note: levrd~=0 predicts success perfectly
      levrd dropped and 200 obs not used

Note: niti~=0 predicts success perfectly
      niti dropped and 12 obs not used

Iteration 0:  Log Likelihood =-147.87523
Iteration 1:  Log Likelihood =-125.56874
Iteration 2:  Log Likelihood =-125.34637
Iteration 3:  Log Likelihood =-125.34629

Probit Estimates                                     Number of obs =     214
                                                     chi2(6)       =   45.06
                                                     Prob > chi2   = 0.0000
Log Likelihood = -125.34629                          Pseudo R2     = 0.1524

------------------------------------------------------------------------------
   innov |    Coef.    Std. Err.      z     P>|z|      [95% Conf. Interval]
---------+--------------------------------------------------------------------
   ismes | -.1866416   .2605415    -0.716   0.474    -.6972935    .3240104
   psmes |  .7530764    .294149     2.560   0.010     .1765549    1.329598
    size |  1.056903   .1889024     5.595   0.000     .6866613    1.427145
   count |  .3188353   .1990749     1.602   0.109    -.0713444    .7090149
  regtyp |  .0988443    .291205     0.339   0.734     -.471907    .6695957
  sector |  .1643506   .1908191     0.861   0.389    -.2096481    .5383492
   _cons | -.9573663   .2804404    -3.414   0.001    -1.507019   -.4077133
------------------------------------------------------------------------------
```

probit levrd ismes psmes niti size count regtyp sector

```
Iteration 0:   Log Likelihood =-294.48678
Iteration 1:   Log Likelihood =-261.42798
Iteration 2:   Log Likelihood =-261.18637
Iteration 3:   Log Likelihood =-261.18633
```

Probit Estimates

```
Number of obs =      426
chi2(7)       =    66.60
Prob > chi2   =   0.0000
```

Log Likelihood = -261.18633

```
Pseudo R2     =   0.1131
```

levrd	Coef.	Std. Err.	z	P>|z|	[95% Conf. Interval]	
ismes	.7451784	.1644704	4.531	0.000	.4228224	1.067534
psmes	-.155014	.1661879	-0.933	0.351	-.4807362	.1707083
niti	.4527946	.2222982	2.037	0.042	.0170982	.888491
size	.2084052	.131434	1.586	0.113	-.0492007	.4660112
count	-.3904393	.1457869	-2.678	0.007	-.6761764	-.1047022
regtyp	-.1790364	.1852699	-0.966	0.334	-.5421588	.184086
sector	.2171681	.1306147	1.663	0.096	-.038832	.4731682
_cons	-.4480533	.1947872	-2.300	0.021	-.8298292	-.0662774

probit ismes niti size count regtyp sector

```
Iteration 0:   Log Likelihood =-288.82075
Iteration 1:   Log Likelihood =-268.12995
Iteration 2:   Log Likelihood =-268.01939
Iteration 3:   Log Likelihood =-268.01937
```

Probit Estimates

```
Number of obs =      426
chi2(5)       =    41.60
Prob > chi2   =   0.0000
```

Log Likelihood = -268.01937

```
Pseudo R2     =   0.0720
```

ismes	Coef.	Std. Err.	z	P>|z|	[95% Conf. Interval]	
niti	.429492	.2338868	1.836	0.066	-.0289178	.8879018
size	.3746418	.127878	2.930	0.003	.1240055	.6252781
count	-.1803166	.1420441	-1.269	0.204	-.4587178	.0980847
regtyp	-.1532309	.1865271	-0.821	0.411	-.5188174	.2123555
sector	.577855	.127207	4.543	0.000	.3285339	.8271762
_cons	-.0829924	.1837685	-0.452	0.652	-.4431722	.2771873

probit psmes niti size count regtyp sector

```
Iteration 0:   Log Likelihood =-266.80463
Iteration 1:   Log Likelihood =-262.23199
Iteration 2:   Log Likelihood =-262.23008
```

Probit Estimates

```
                                          Number of obs =      426
                                          chi2(5)       =     9.15
                                          Prob > chi2   =   0.1033
Log Likelihood = -262.23008               Pseudo R2     =   0.0171
```

psmes	Coef.	Std. Err.	z	P>|z|	[95% Conf. Interval]	
niti	.1738437	.2132073	0.815	0.415	-.2440349	.5917222
size	-.0484944	.1295515	-0.374	0.708	-.3024106	.2054218
count	-.014127	.1460805	-0.097	0.923	-.3004396	.2721856
regtyp	-.2230986	.1801641	-1.238	0.216	-.5762138	.1300167
sector	.3137847	.1280011	2.451	0.014	.0629072	.5646622
_cons	-.4293507	.1825892	-2.351	0.019	-.7872189	-.0714825

probit niti ismes psmes count size sector regtyp

```
Iteration 0:   Log Likelihood = -137.1623
Iteration 1:   Log Likelihood =-126.36177
Iteration 2:   Log Likelihood =-125.79383
Iteration 3:   Log Likelihood = -125.7866
Iteration 4:   Log Likelihood = -125.7866
```

Probit Estimates

```
                                          Number of obs =      426
                                          chi2(6)       =    22.75
                                          Prob > chi2   =   0.0009
Log Likelihood = -125.7866                Pseudo R2     =   0.0829
```

niti	Coef.	Std. Err.	z	P>|z|	[95% Conf. Interval]	
ismes	.3682405	.2294234	1.605	0.108	-.0814211	.8179022
psmes	-.0267771	.2112848	-0.127	0.899	-.4408878	.3873335
count	-.7876322	.2447246	-3.218	0.001	-1.267284	-.3079808
size	.2785939	.1847363	1.508	0.132	-.0834826	.6406704
sector	-.0306546	.1806659	-0.170	0.865	-.3847532	.3234441
regtyp	.1202182	.2312529	0.520	0.603	-.3330291	.5734655
_cons	-1.586852	.2762283	-5.745	0.000	-2.12825	-1.045455

Druck: Strauss Offsetdruck, Mörlenbach
Verarbeitung: Schäffer, Grünstadt